NEGOTIATING
CULTURAL
ENCOUNTERS

NEGOTIATING CULTURAL ENCOUNTERS
Narrating Intercultural Engineering and Technical Communication

Edited by

Han Yu
Gerald Savage

Professional
Communication
Society

IEEE PRESS

A JOHN WILEY & SONS, INC., PUBLICATION

Library of Congress Cataloging-in-Publication Data:

Negotiating cultural encounters : narrating intercultural engineering and technical communication / edited by Han Yu, Gerald Savage.
   pages cm
Includes bibliographical references.
ISBN 978-1-118-06161-9 (pbk.)
1. Communication of technical information. 2. Intercultural communication. I. Yu, Han, 1980– II. Savage, Gerald J.
   T10.5.N44 2013
   303.48'2—dc23   2012030895

Printed in the United States of America.

10 9 8 7 6 5 4 3 2 1

CONTENTS

THEME-BASED TABLE OF CONTENTS

Knowledge and Technology Transfer

PREFACE

Although intercultural communication has become an important area of research, practice, and teaching in the United States in the last 20 years or so, we have not heard many concrete stories on how intercultural technical communication is practiced in everyday workplaces. Such stories are usually tucked away in individual practitioners' memories or shared with a few colleagues and friends over lunch or coffee. These stories, we believe, have value and must be told and studied in order to advance the research, practice, and teaching of intercultural technical communication.

Storytelling is a fundamental means through which human beings form identities, build knowledge, and handle human interactions. Telling and listening to stories is therefore an important way for people to form and understand their cultural identities. Allowing practitioners to tell personal stories also is a form of empowerment. By narrating their lived intercultural experience—the good, the bad, and the ugly—practitioners can share their lessons learned with each other and educate fellow practitioners and management who are not experienced in intercultural projects. These stories also relay the lore of practice, which almost always emerges in advance of formal research or codified knowledge. In the same way, stories provide opportunities to test current theories and knowledge paradigms.

Pedagogically, these narratives are instrumental for teaching intercultural competencies to students who do not, yet, have their own intercultural stories to draw on. In these cases, reading other people's stories provides the best vicarious learning. Personal narratives contain concrete details, contextualized contradictions, and individual choices. Confronted by these, students will learn to see culture not as a monolithic entity that can be easily codified, but instead as complex, fluid, and personal realities that must be carefully and continuously negotiated.

Negotiating Cultural Encounters: Narrating Intercultural Engineering and Technical Communication fulfills these goals. This book includes 12 narratives from 13 technical communication professionals. These authors represent six nationalities, some with multiple cultural identities or experiences. Their stories portray interactions with people and organizations from at least 19 cultures or nationalities and cover a range of communication issues in at least ten different technical fields. Most of these stories are based on the authors' own experience, and a few are based on the experience of fellow colleagues. Some authors choose to tell their own experience in first-person narratives and others prefer to distance themselves and tell their own stories in the third person. Some stories reveal the complexities of working with globally dispersed teams, some focus on working with colocated teams who have diverse cultural backgrounds, and others focus on working with "domestic" teams that nevertheless include people of various professional, organizational, and regional cultural identities. Each story comes with discussion questions that can be used either as a self-study guide or for class discussion. Additional readings are also suggested at the end of each story for readers interested in learning more about particular topics.

The narratives in this book sometimes confirm but often challenge common notions of what working across cultural and linguistic boundaries involves in professional communication settings. They help readers discover the point at which analysis reaches the limit of its usefulness because the unique time, place, and circumstances of each story do not yield a solution based on cultural rubrics or heuristics. We see that people in such situations have to make judgments, take risks, weigh consequences for various possible actions, and learn to negotiate cultural encounters on their own.

Some of the stories have happy endings and feature culturally sensitive characters; others end with unresolved issues and reveal characters who may have narrow or biased worldviews. None of them, however, is simple or straightforward. Together, these stories make it clear to any reader that culture is never finished; it isn't "something" for which you can provide a definition or theory and make it fit every intercultural situation. Negotiating cultural encounters requires sincere and deep-seated cultural sensitivity, a willingness and even eagerness to learn alternative practices and solutions, and the ability to always find new footing, consider contextual factors, question taken-for-granted knowledge, and never cease paying attention.

HAN YU
GERALD SAVAGE

A NOTE FROM
THE SERIES EDITOR

The IEEE Professional Communication Society (PCS), with Wiley-IEEE Press, is excited to continue its book series titled *Professional Engineering Communication* with this collection of insightful stories about the struggle for cross-cultural communication. Bringing together technical communication experts from around the globe, editors Han Yu and Gerald Savage have invited these authors to share their insights and experiences into the rewarding and sometimes frustrating important work of communicating across cultural borders. And while the authors reveal to us that, in the end, success and understanding can be built, we learn by reading their tales of how hard-won those successes can be. Someone once told me, "The technical work is easy. It is the people that are hard," and that sentiment is beautifully reflected here with voices many and varied. We can learn much from their on-site and very real challenges to communicate with purpose, efficiency, strategy, grace, and professionalism.

As you read, you will encounter many Englishes. While some people turn to American Standard English or British Standard English as a baseline, the goal of this book is also to reveal the complexity of working with other Englishes, gaining respect for the incredible work that gets done by translators, localization experts, documentation managers, editors, and technical communicators worldwide. Certainly, in these pages, you will find a turn of phrase or usage that is not familiar. It is in these slightly uncomfortable moments that we all can learn just a bit more about how English, as the language of technical work, is a fluid and accommodating element that surrounds and embeds itself in our very perceptions of the world.

I want to back up a bit and talk about this new PCS-sponsored project. As a series, *Professional Engineering Communication* (PEC) has a mandate to explore areas of communication practices and application as applied to the engineering, technical, and scientific professions. Covering the realms of business, governmental agencies, academia, and other areas, this series will develop perspectives about the state of communication issues and potential solutions when at all possible. This book belongs squarely in this series because it discusses the essential work of communicating engineering, technical, and business ideas across languages, political lines, custom, and even generations.

The books in the PEC series will keep a steady eye on the applicable while acknowledging the contributions that analysis, research, and theory can provide to these efforts. Active synthesis between on-site realities and research will come together in the pages of this book as well as other books to come. There is a strong commitment from PCS, IEEE, and Wiley to produce a set of information and resources that can be carried directly into engineering firms, technology organizations, and academia alike.

We invite you to dive into these stories, where we come to understand how communication is never simple and never predictable. It is in those struggles that moments of clarity and purpose come into a stronger light, and the authors want to share those moments with you.

TRACI NATHANS-KELLY, PH.D.

INTRODUCTION: A STORY OF STORIES

Han Yu and Gerald Savage

Han Yu is an associate professor in the English Department, Kansas State University. Her research focuses on writing assessment, intercultural technical communication, and visual communication. Han's work has appeared in Business Communication Quarterly, IEEE Transactions on Professional Communication, Journal of Business and Technical Communication, Journal of Technical Writing and Communication, Programmatic Perspectives, Technical Communication, *and* Technical Communication Quarterly. *Han teaches classes in technical communication, business writing, engineering writing, and science writing. She worked as a technical editor for State Farm Insurance and as an editor and translator for the New Oriental Publishing Group in Beijing, China.*

Jerry Savage is a professor emeritus of Rhetoric and Technical Communication at Illinois State University. His research focuses on diversity and social justice issues in international and intercultural contexts of technical communication practice. He is co-editor with Dale Sullivan of Writing a Professional Life: Stories of Technical Communicators On and Off the Job *and with* Teresa Kynell Hunt of Power and Legitimacy in Technical Communication, *Volumes 1 and 2. His work has appeared in numerous books and professional journals. He is a fellow of the Association of Teachers of Technical Communication and has been awarded the Jay R. Gould Award for Excellence in Teaching Technical Communication by the Society for Technical Communication and the Distinguished Service Award by the Council for Programs in*

Technical and Scientific Communication. In addition to teaching, he worked for many years in technical, management, editorial, and writing roles in business, industry, and government.

A PROLOGUE FROM HAN: WHERE IT ALL STARTED

The idea for this collection came from a graduate class that I regularly teach at Kansas State University. The course, Introduction to Technical Communication, is typically filled with students who are not technical communication majors (at the time of this writing, my department does not offer a technical communication track for graduate students but offers courses toward a certificate). These students come from all backgrounds: literature, creative writing, rhetoric and composition, modern languages, and communication studies. Rarely do I have a student who has technical communication work experience. Many of them choose to take the class because they want to know more about technical communication or because they believe that technical communication is about gaining "practical" skills such as writing "technical" manuals which may help improve their odds for employment upon graduation.

This scenario should not sound unfamiliar to many technical communication teachers. As a young, interdisciplinary, and growing field, technical communication seems to defy definition, as Jo Allen argued in "The case against defining technical writing" back in 1990 [1]. What Allen wrote made sense then and still makes epistemological sense today, but to tell students that "we cannot define what technical communication is" in a class that is supposed to introduce the discipline just won't do. So I have my students read the field's landmark works and learn about its major issues, from Connors' historic study [2] to Miller's rhetorical turn [3], from the discussion of audience to ethics, visuals, and usability.

These readings are helpful for students to form a theoretical and conceptual understanding of the field, but they do not give students a realistic sense of what technical communicators do day in and day out on the job and how theories and concepts play out in the real workplace. Ideally, students should gain this realistic understanding through internships or other empirical experiences such as client projects, but these are unfortunately hard to do (at least do well) in an introduction class or even through a short certificate program.

In the absence of actual experiences, stories provide an excellent alternative for supplying empirical knowledge. As scholars in anthropology, psychology, and sciences have argued, stories (or narratives) are a primary thought and discourse process for human beings to form identities, make sense of experiences, develop rationality, build knowledge, and handle human interactions [4–6]. The hunting expeditions and other events painted on cave walls from prehistoric times are the cave dwellers' attempt to, among other things, preserve the history and identity of their communities. In oral cultures, story-telling is a powerful way for people to obtain knowledge and sustain tradition. Even in print cultures, stories retain their visceral power in human communications. Fundraising campaigns, for instance, often use stories of individuals in need to evoke empathy and inspire philanthropy. And the act of reading stories is believed to strengthen one's rationality and ability to form argument because of our constant habit of testing narrative fidelity against lived experiences [5].

But the field of technical communication, Perkins and Blyler argued, has a rather low regard for stories and narrations [6]. This attitude, Perkins and Blyler speculated, has to do with "the Western tradition of privileging logic and science" and "the linking of professional communication to an objective view of rhetoric" [6, p. 11]. Logic and science are regarded as fundamental truth independent of human subjectivity; stories, because they are based on personal experiences, belong to literary genres such as novels and epics but are not objective enough for technical and professional communication. Narratives are seldom mentioned in textbooks. When they are, they are recommended only for supposedly low-level and uncritical writings such as recording chronological events or catching readers' attention [6].

A notable exception is Savage and Sullivan's 2001 collection *Writing a Professional Life: Stories of Technical Communicators On and Off the Job* [7]. A first of its kind in the field, Savage and Sullivan's collection has practicing technical communicators tell stories about their work and life. Together, these stories give a sampling of what it is like to be a professional technical communicator in the United States at the turn of the twenty-first century: what they do, whom they work for and work with, how they get their education and training, and the ups and downs of their work and life. Advice and lessons learned, some explicit and others implicit, are offered.

My students enjoy reading these stories. As they put it, these stories help them to experience, vicariously, what it is like to be a technical communicator and to see whether that kind of work/life is what they would want. I like assigning these stories too. They provide lively class discussion and are much easier to arrange than inviting guest speakers to share narratives in person.

There is, however, one thing that I wish Savage and Sullivan's collection could have done, which is to include stories about intercultural technical communication. An increasingly important topic in the field, intercultural technical communication is now being taught in a variety of classes at various levels. In my Introduction to Technical Communication class, for instance, the class discusses the importance for today's technical communicators to reach international audiences and to work with colleagues from diverse cultural backgrounds, reads about the communication needs and rhetorical preferences of audiences from different parts of the world, and discusses the tools, processes, and challenges of intercultural technical communication.

Although intercultural narratives did not find their way into Savage and Sullivan, I did find something else: intercultural technical communication case studies. Case studies have been widely used in educational settings, especially in business and law. They are "designed to move business students from the abstract to the specific by forcing them to grapple with realistic details and situations ... to provide mock contexts in which students can practice their legal arguments" [8, p. 21]. Case studies have also become a familiar teaching method in technical communication and regularly appear in textbooks such as Markel's *Technical Communication* [9]. For instance, a case on ethical communication may describe an ethical dilemma faced by a technical communicator and ask students to put themselves in the communicator's place to try to solve the dilemma. Johnson-Sheehan and Flood call these cases the "closed cases" [8]. They are "closed" because "the rhetorical situation for the case is almost wholly contained within the assignment itself.... an enclosed technical setting is described for the students: a problem or issue is identified, and the students are given roles to play.... students are asked to respond within a set of guidelines that are wholly prescribed by the teacher" [8, p. 21].

Similar case studies can be found that address issues in intercultural technical communication—although they are few and far between. One title stands out: Deborah Bosley's 2001 collection *Global Contexts: Case Studies in International Technical Communication* [10]. This collection contains 12 cases, from developing technical documentation with Taiwanese partners, to adapting to South American communication patterns, to building a cyber university with a Malaysian company. Each case is consistently presented: "background information, to place the case in context; a description of the problem and the people or agencies involved, including pertinent graphics and communication examples; background for analysis, which introduces relevant theory, such as Edward Hall's concept of high- and low-context cultures or Lawrence Kohlberg's schema for cognitive moral development; and a brief analysis" [11, p. 128].

Bosley's cases are considerably longer and much more complex than the typical closed cases. But having used them with my students, I started to see their similarities: in Johnson-Sheehan and Flood's words, "an after-the-fact detective game in which given clues lead students to discover who or what did it" [8, p. 22]. The consistent format of the cases, in particular, helps to create this impression: The "background of the case" section confronts students with a baffling communication problem; the "background for analysis" then provides students with the clues that they need to unravel the mystery; and "analysis of the case" walks students through the process, putting the different clues in place.

No doubt, when my students follow these sections and read through the cases, they encounter valuable intercultural communication scenarios, and they gain important knowledge and problem-solving skills, but as the teacher, I wish these scenarios could be less like closed case studies and more like open stories. Case studies and stories share many similarities. To start, they both contain scenarios, characters, and exigencies, and both can use contextual details to engage and educate readers. But they also have important differences. Case studies, with their origin in the structured and regulated fields of business and law, are expected to offer some, if not always definitive, answers. Their focus is on analytical learning: helping students to piece together clues toward those answers. Stories, with their origin in everyday human life, are more fluid, unique, complex, and open to individual interpretation. They offer opportunities for students to develop critical learning: presenting problems and engaging readers to think about those problems and explore their own answers.

Both case studies and stories are important pedagogical tools in different contexts. Stories can be more instructive for teaching intercultural communication. This is because cultural issues are almost always complicated and defy a neat list of clues—they are more like a web of multiple, fluid, and interwoven factors. What's more, how we negotiate those issues and interpret those factors is influenced by our own cultural bias, rendering it impossible for us to arrive at set answers—who, after all, do we exclude when we say "we" have the answer? An example should illustrate my point.

Because of my Chinese background, I have been asked, often enough, by students in my intercultural technical communication class and by curious colleagues about the preferred color choices for communicating with Chinese people. Having gleaned some information from cultural handbooks or online and anecdotal sources, they want to confirm if white is, indeed, a "bad" color for the Chinese because it is the color of mourning in China whereas red is a "good" color because it is the color of festivity.

Every time, I begin my answer with "No, it is not that simple." This kind of response, I suspect, disappoints some of my questioners who want a straightforward answer or expect, consciously or unconsciously, an answer that reinforces their idea of China as a different culture where interesting rules govern such matters as color choices.

White and red, I continue to explain to my questioners, are used in different contexts in China—as everywhere else. White can indeed be a symbol of mourning (and "bad" in that sense) because traditional Chinese funerals, referred to as the "White Event," use white as a primary color. The blood relatives of the deceased will dress in white, women will wear flowers made of white paper, and white fabric is used to decorate the service area. However, black is also an important color in Chinese funerals: It is worn by non–blood relatives, friends, and other attendees of the event to show respect and soberness. Because of the influence of Western traditions, black is even replacing white as a dominant color in contemporary Chinese funerals, especially in urban areas. To add to the twist, white can also be a primary color at an important celebratory event (and "good" in that sense) in China: weddings. Influenced by Western traditions, contemporary Chinese weddings, especially in urban settings, feature white bridal dresses, white flowers, and white decors to symbolize purity and elegance.

The color red is an equally complex matter. Celebratory events such as weddings and birthdays in China are traditionally known as the "Red Event" and red is the dominant color used at those events. White dresses may have become the default choice for China's urban brides, but many of them, for parts of their reception ceremonies, will change into traditional Chinese Qipao, a dress made of red fabric in various shades and with various patterns. At traditional celebratory events, the Chinese characters "joy" (喜) and "happiness" (福) are cut out in red paper and pasted on walls and doors, red lanterns are raised, red-colored firecrackers are played, and red envelopes are used by parents and friends to give money to children. So, indeed, red is often associated with joy and prosperity (and "good" in that sense). But the same bright red can also have very different meanings: It symbolizes blood, and in the political context, blood that was shed by the Chinese people and troops in founding and defending the People's Republic of China. The emblems of the Chinese Communist Party and the People's Liberation Army and the national flag of China are predominantly red. Red therefore represents the Party, the Army, and the China that we know today. For people who are devoted to the Party, red is a "good" color that suggests patriotism; for people who are not, it can be a "bad" color associated with suppression and bloodshed of an altogether different type.

If the use of two colors in one culture can be so complex (and this is only a glimpse into that complexity), we can only imagine the amount of intricacies our communicators face in intercultural encounters that are associated with a variety of characters, events, deliverables, timelines, technologies, and numerous other factors. To try to present all of these as clues from which some kind of answers will always emerge seems idealistic at best. Instead, a mode of learning and teaching that is less prescriptive and more exploratory and less teacher centered and more learner centered seems more appropriate. Open-ended stories about intercultural technical communication can help realize this education. Through stories, students can see and hear, from practicing communicators, what intercultural technical communication is all about: what tasks do practitioners work on, whom do they work with, what issues do they encounter, and what processes and skills are involved. With these stories, students are expected to think on their own, to raise questions, to suspect assumptions, and to debate possible solutions; they will need to juggle logic, probability, ethics, justice, creativity, and even an attempt at wisdom. A good narrative will provide a situation to which students and readers can apply the intercultural communication principles and concepts we want them to know analytically. The narrative will also help them discover the point at which analysis reaches the limit of its usefulness because the situatedness of the story will usually involve some key aspect that is unique and does not yield a solution based on analytic terms. That's where they will have to make judgments, take risks, weigh consequences for various possible actions, and learn to negotiate cultural encounters on their own. Essentially, that is the real decision making our students and practitioners will have to make in real-life intercultural communication.

In addition to these pedagogical effects, simply allowing practitioners to tell their stories is significant. As Perkins and Blyler state, telling stories about what we do can empower our profession [6]. Although intercultural technical communication has become an important part of our field's research and teaching, it is nonetheless still an emerging topic and demands more innovative research, meaningful pedagogies, and research-based attention in industry. By telling stories about their work, their lives, their frustration, and their accomplishments, our intercultural technical communicators can have their voices heard. Through workplace narratives, intercultural professional communicators relay the lore of practice, which almost always emerges well in advance of formal research, theory, or codified knowledge. Such narratives provide opportunities to test current theories and knowledge paradigms and suggest directions for new research. By following these narratives, managers, professionals in related fields, students in training, and educators and researchers can hope to develop more effective and appropriate intercultural technical communication.

To sum it up, in early 2010, I started to consider putting together a collection of narratives of intercultural technical communication. I approached Gerald (Jerry) Savage with this idea, who enthusiastically agreed to join me. From there, we started talking, face to face, on the phone, and through lengthy e-mails, to hash out the best way to realize my ideas. By the summer of 2010, we had a plan of what we wanted.

A STORY OF STORIES FROM HAN AND JERRY: A LEARNING PROCESS

We Thought We Planned It All

First, we wanted to hear from practicing technical communicators who can share the most pertinent and current stories. It is true that many university educators and intercultural scholars have intercultural experience and have valuable stories to tell, but they also have more venues to have their voice heard: journals, conventional textbooks, and previous case study collections such as Bosley's. Practicing technical communicators, on the other hand, have fewer opportunities (beyond such venues as conferences and trade publications) to reach students, teachers, researchers, and fellow practitioners; many of them also lack the time to do the reaching-out beyond their professional duties. So engaging these practitioners (without excluding researchers and educators) can help us get fresh perspectives and essential information that we may otherwise never realize existed.

Second, we hoped that our authors would have diverse backgrounds in terms of their cultural identities, professional affiliations, workplace roles, and communication assignments so their stories could demonstrate a variety of scenarios, challenges, possibilities, and complexities. Current intercultural technical communication publications, especially textbooks, are predominantly authored by U.S. writers or those who reside in the United States. As a result, the intercultural interactions we read the most are those between U.S. culture and "other" cultures—and primarily from the U.S. perspective of how to "cope with" those cultures. While these interactions and perspectives *are* important for our students and readers to know, focusing *only* on those is a rather instrumental, even ethnocentric, way of learning intercultural communication. So we hope our collection can give students and readers a chance to look beyond the U.S. worldviews and glean some non-U.S. perspectives.

Third, we wanted concrete narratives, not general overviews or descriptions of our authors' work. The familiar 5Ws and 1H are the guidelines we gave them: Who was involved? What happened? Where did it take place? When did it take place? Why did it happen? And How did it happen? Jerry also coined a SCAD system to coach our writers:

- *S for Situation.* The occasion that makes the story. This may be an incident, a crisis, or a typical occasion that illustrates what it's like to be an intercultural technical communicator.
- *C for Character.* People, especially the main characters. These should be real human beings with distinctive nature, character, personality, temperament, and so on.
- *A for Action.* Important, significant, or interesting things that happen in the story.
- *D for Dialogue.* Verbal communication as well as other modes of communication.

We insisted on concrete stories with rich details so we could provide the contexts that are essential for students and readers to meaningfully learn about culture and intercultural interactions. These details help the book to stay away from uncritical instruction of cultural heuristics such as those developed through studies by Edward T. Hall, Geert Hofstede, and Trompenaars and Hampden-Turner. These heuristics were widely taught and include such well-known dimensions as low context–high context culture and individualism–collectivism. But an uncritical instruction of cultural heuristics tries to use preestablished dimensions to map out the characteristics of a culture and ignores contextual details that influence intercultural communication.

In his essay, "Culture and cultural identity in intercultural technical communication," Hunsinger pointed out various problems with such an approach—although he did not differentiate critical and uncritical uses of heuristics and argued against the heuristic approach in general [12]. Operating from the heuristics alone, one may ignore extracultural elements (political, economic, historical, technological, and personal ones) that shape intercultural interactions—an act of oversimplification and essentialism [12]. Furthermore, because it is tempting to read heuristic factors as binaries (e.g., individualistic vs. collectivistic) even when they are meant to describe tendencies, the heuristic framework can perpetuate differences or lead some to believe that "our" way is better than "their" way—an act of ethnocentrism.

To avoid these difficulties, as Hunsinger suggested, drawing from Arjun Appadurai, we planned to distinguish "culture" and the "cultural" in our collection of narratives [12]. According to Appadurai's critical cultural theory, *culture* as a noun describes some monolithic and definitive substance, whereas *cultural* as an adjective (as in "cultural identity") is situated and embodied [13]. When we do not sufficiently separate the two, we inevitably assume that cultures (which we have confined, a priori, to static heuristic descriptors) determine the fluid and multiple cultural [13]. Appadurai argued that this assumed causal relationship should be reversed: "rather than trying to identify the ways culture manifests itself in the cultural, Appadurai studied the ways the idea of culture is constructed from the apparently cultural" [12, p. 38].

By insisting on concrete stories and rich details, we hope this book will achieve the same thing: *Rather than asking that students and readers remember static heuristics, we want to show them how cultural performance shapes our understanding of culture.* We want readers to see what happens in intercultural interactions; what is said or done and by whom; what are people's motives, concerns, and emotions; what arguments break out; what compromises are made and how they are made; and what solutions are reached or not reached. We hope that by reading these apparently cultural narratives, students will learn to critically approach culture and cultural heuristics.

Last, we planned to refrain from interjecting our own interpretations and explanations of the narratives. We want our writers to share their perspectives, beliefs, and solutions of cultural encounters, but we will neither endorse nor prescribe their views as if they are the "right" ones. Each writer is a cultural agent so their understandings are necessarily influenced by their cultural backgrounds and upbringings, and yet, they are also able to learn and move beyond their prior culturally constrained viewpoints. For readers—who are also cultural agents—it is important that they learn to negotiate cultural encounters on their own: to read the stories presented to them, ask questions, examine the writers' biases, examine their own biases, and eventually get at a more sophisticated way of thinking about and approaching cultural encounters.

Without set answers, our students and readers may feel lost or they may question their own ability to arrive at the right answers. But once again, there is no right answer and getting answers is not the goal of the book. We want readers to accept the complexity, ambiguity, and contradictions that are inevitable in intercultural interactions and not expect that a book (or a college degree or a certain level of experience) will ever get them to a point where they have all the knowledge necessary to communicate across cultures without misunderstandings or difficulties in a particular—and always in some sense new—situation. The best we can do in a book or class is to engage students to critically think about cultures, to problem-solve cultural issues, to participate in intercultural communications, and to continue this process even after they leave the book or class behind. Certainly, *some* instructional parameters are needed to guide students and readers in their learning, so we decided to add discussion prompts and questions to each story and suggest additional readings when that is appropriate. We also decided to use an additional, theme-based table of contents that groups the stories into broad categories to help teachers and readers see various commonalities and connections.

We Realized We Didn't Know It All

Having firmed up our plans, we set to work. For a while, things were going well. By the end of 2010, we had identified a group of excellent writers whose profiles fit what we were looking for. Most of them are industry practitioners, including freelancers, consultants, and corporate employees. We also have a few educators who have experience in industry, or who have close contacts in industry whose experiences became the basis of their narratives. Our authors have diverse cultural identities and backgrounds, as their biographies show. Their stories also cover a range of communication issues in technical, engineering, and related fields, including web design, documentation management, translation, developmental projects, workplace safety, business presentations, and more.

Some stories reveal the complexities of working with globally dispersed teams, others focus on working with colocated teams who come from different national cultures, and still others focus on working with "domestic" teams who nevertheless have diverse cultural identities. We are very glad to include this last type of "cultural encounters" because national cultures and international contexts should not be the default anchor point in the studies of cultures—although they often are in current literature and textbooks. Within nation states, there can be various subcultures based on geographic locations, social/economic statuses, professional circles, and other factors. We only wish we could have more of such domestic cultural narratives. But then, again, we wish we could have more of every type of narratives, since there is simply no end to the complexities in intercultural interactions—but a book has to end somewhere.

In the early months of 2011, initial drafts started to trickle in from our writers. We would read drafts separately, add side comments and queries, make tracked changes, and write up summaries of major questions or concerns. While these are common processes we had expected to go through as editors, we quickly realized that we had not planned for everything. And that revelation, clichéd as it may sound, was a true learning experience for both of us—and for our writers too, as many of them tell us.

We had, from earlier on, made the plan not to present our writers' cultural perspectives and solutions as if they are the right ones or to interject our own interpretations of the stories in our editing process. But we conveniently overlooked the fact that with the array of cultural and professional backgrounds of our writers, ourselves, and our potential readers, things are never going to be that easy. Let's consider a hypothetical example: A U.S. writer writes about a U.S. manager who was parachuted into a Chinese company, was all wise and confident, offered his Western experiences, triumphed over all the Chinese employees, and rescued the company's faulty documentation practices. With such a story, we can justifiably be concerned about sending a "neo-colonial" message or at least be concerned that some readers will see it that way. But what if the writer is Ghanaian, whose nation was subject to centuries of colonialism? Wouldn't that make a difference, if just an unconscious one, in how our readers see the story? And what if the protagonist in the story was a Ghanaian manager? Shouldn't we consider how Africa's colonial history might have shaped his cultural identity and how he conducted himself in a foreign country? Then what if our Ghanaian writer and/or manager turned out to be Western educated and employed—westernized, as we often say? And so the complexities continue.

Eventually, we learned to accept what, in retrospect, seems so obvious (but then again, nothing in cultural learning *is* obvious): To even try to block out our cultural identities or those of our writers from the narratives is simply not possible. As soon as we put pens to paper and start writing or editing, we are allowing our cultural selves to guide what we say and how we say them. We want to make this inconvenient truth clear to our readers *so they can feel free to "judge" the stories but at the same time remember to judge themselves,* for the readers are cultural agents just as we are, and they should be aware of the biases they bring to the interpretation of the narratives. Ultimately, this is what intercultural interactions are about: None of us can or should erase our cultural identities, but we can and should be mindful of those identities and how they are projected in intercultural interactions. As Geertz put it, the central question is not "how to prevent subjective views from coloring objective facts," but "how best to get an honest story honestly told" [14, p. 9].

How, then, can we get these honest stories honestly told? A distinction needed to be made between two voices in the story: the voice of the protagonist and the voice of the story author. The protagonist is free to present to the world his or her cultural perspectives, beliefs, and solutions, but the author does not necessarily share the views of the protagonist—even if in many situations the two are one and the same. This distinction is rather significant when we think about the Foucauldian concept of author.

As Foucault argued, author, or what he calls the "author function," is not an individual but arises from and comes to stand for a complex set of social operations and constraints that determine how we interpret text: "the comparisons we make, the traits we extract as pertinent, the continuities we assign, or the exclusions we practice" [15, p. 127]. This author or author function is not the protagonist or the individual narrator of a story. Given a novel that is narrated in the first person, it is easy for readers to remember that the protagonist "I" is not the author but a "'second self' whose similarity to the author is never fixed and undergoes considerable alteration within the course of a single book" [15, p. 129]. With this tacit understanding, there is less danger of readers confusing a protagonist's (possibly troublingly narrow and biased) worldview with what the author endorses. But given our narratives, which are presented as real-life experiences, readers will be tempted to equate the protagonist "I" with the authorial voice of a story and even that of the book as a whole. At the same time, the other characters in the narrative, because they are not "authors," have no authority to tell *their* versions of the story so their emotions, concerns, and beliefs are excluded.

We shared this concern with our writers and asked that they do the impossible task of separating their personal voice from their authorial voice—And so they did. They include more self-reflections in their stories to create the Foucauldian "alter ego"; they candidly and bravely admit the protagonist I's (which was often their own) helplessness, doubt, ignorance, and mistakes; and they avoid applying cultural labels to other characters in the story but instead tell what they did or said, allowing readers to draw their own conclusions.

Presenting the "apparently cultural," as we planned, also proved difficult because everyone's understanding of what is *apparently* cultural is different. At the earlier stage of the editing, we often prompted our writers to give concrete details about the cultural incongruities, differences, or even conflicts that pertain to their stories. We were not trying to exaggerate the performed cultural differences. Rather, we thought that our writers, being immersed, day in and day out, in their intercultural environments, might have become so used to the diversities and challenges of those environments that they overlooked details that would be educational to our readers. But gradually, we started to wonder whether, in our attempt to create learning moments, we were doing nothing but emphasizing differences that we thought "mattered"— when in some situations it is not at all obvious whether communication challenges are based on cultural factors or on interpersonal, personal, stress-related, task-related, or any other possible factors. Nor is it obvious that the interpersonal, personal, stress related, or task related is not cultural.

In addition, as some of our writers reminded us, although cultural differences do exist, today's globalized economy, communication networks, and widespread diaspora are creating a growing number of working professionals who share similar work habits and professional standards. The situation of Europe is a telling example. As our writers tell us, industry standardization across Europe means that people share a similar technological infrastructure, working practices, and organizational structures. In particular, the emergence of the European Union brought cultures across Europe much closer. Peoples can freely move across country borders to live and work, and organizations and businesses across Europe also conform to basic standards.

At that point, we realized (again, what seems clear in retrospect) that our focus should not be on what "cultural differences" exist but, as Kubota and Lehner suggested, we needed to move "toward questions such as: *How have we come to believe that a certain cultural difference is true?, What political purposes have motivated the construction of particular beliefs about cultural difference?, and What alternative understandings of cultural difference, or counter-discourses, are available to transform our taken-for-granted knowledge?"* [16, p. 17].

There is no one-size-fits-all answer to these questions when it comes to our stories. Each time something came up, we pondered, questioned our own assumptions, and negotiated for tentative solutions. We would then send lengthy comments to writers, explaining our concerns, inviting their feedback, and suggesting possible revisions. We continued these conversations with each writer, we debated, and we made compromises, until all felt relatively comfortable with the end product. We cannot say that we are convinced every decision we made is perfect, but we made the best decisions we know on how to get honest stories honestly told.

A Note on Styles and World Englishes

Readers who are sensitive to stylistic issues will notice inconsistencies in style from chapter to chapter in this collection. Here, we refer to style in the sense of spelling, punctuation, and some aspects of syntax. Some of these inconsistencies may well appear to be errors and oversights in the copyediting process. But, in fact, they reflect different cultural, language, and national conventions of style. As we began reading first drafts of the stories, we found ourselves "correcting" a number of stylistic elements, making them conform to U.S. style conventions. Quickly, however, we realized the ironic contradiction, in a book about cultural diversity, for the editors to attempt to homogenize the styles used by writers from diverse cultural backgrounds and who write in diverse cultural contexts using different stylistic conventions.

So, instead, we contacted the editorial staff at Wiley/IEEE Press to ask that we be allowed sufficient latitude to keep the stylistic conventions and choices to which our writers are accustomed. We recognize the potential for stylistic chaos in this decision and we have agreed to some degree of compromise. We have attempted to distinguish between errors (style elements the writers did not intend and would want to correct if they caught it themselves) and choices based on local conventions.

The choices we have elected to keep in the text include syntactic choices and word choices that may seem like errors to some readers but which represent the English spoken in some nations. For instance, unlike U.S. English, British English tends not to use serial commas (the last comma placed before "and" or "or" in a series). In U.S. English, collective nouns that refer to a group of people (such as "team" or "company") often take on singular verbs or pronouns, but in British English, both singular and plural forms are readily used (e.g., "The company was formed in 1993. They have attracted many talented people over the years"). Whether to use the singular or the plural forms depends on whether one thinks of the group as a singular, impersonal unit or a collection of individuals.

Our decision to accommodate multiple styles is based on studies in the emerging linguistic and rhetorical field of world Englishes. The field of world Englishes is interested in the ways various regions of the world have appropriated or inherited the English language, often from colonial occupation but also as a *lingua franca* used in global trade or diplomacy. A key argument of world Englishes studies is that no nation or culture is the sole proprietor of a language. English is not the exclusive property of the United Kingdom, the United States, Canada, New Zealand, or Australia. It is widely spoken in countries such as India, Ghana, South Africa, Singapore, and many other nations and regions of the world. However, it is spoken in ways specific to those locations, reflecting the indigenous language and cultural conventions of those places and picking up many words and expressions from those local languages and customs. Thus, Indian English may seem "foreign," exotically structured and accented, to a speaker of South African English, and vice versa. These Englishes or, as Wierzbicka calls them, "post-Englishes, are sometimes given hybrid names: Hinglish [Indian English], Singlish (from 'Singapore English' or 'Sri Lankan English'), Pringlish (from 'Puerto Rican English'), and so on" [17, p. 305].

Although the topic of world Englishes is in a sense relevant to all stories in international technical communication, it appears to us to be central in a number of our narratives, which we have categorized under world Englishes in our theme-based table of contents.

A Note on Translation and Localization

Readers who are familiar with the concepts of translation and localization will notice that we combined these two in our theme-based table of contents. This may again appear to some as an error or an editorial oversight, but in fact, it reflects the reality we encountered in our learning process. Although in academic and classroom discussions, we often separate these two concepts, it appears that a number of our industry practitioners use them interchangeably.

Translation, an important stage in producing international technical communication, is about putting documents into the languages of users in various parts of the world. A basic strategy for translation is for the technical writer to compose the documentation in English in a way that makes it translation ready. This usually involves simplifying the language by means of limited vocabulary and eliminating complex syntactic structures. This helps to avoid mistranslation and to help the translation go faster. It may also reduce errors in machine translation or machine-assisted translation.

Localization, another important stage in producing international technical communication, is harder to define, partly because it is a relatively new concept within the field of technical communication. Localization is defined by the Globalization and Localization Association (GALA) as follows [18]:

Describes the process of adapting a product to a specific international language or culture so that it seems natural to that particular region, which includes translation, but goes much farther. True localization considers language, culture, customs, technical and other characteristics of the target locale. While it frequently involves changes to the software writing system, it may also change the keyboard usage, fonts, date, time and monetary formats. Graphics, colors and sound effects also need to be culturally appropriate.

GALA's definition of localization is obviously specific to the interests of many of their member organizations—generally the software industry—however, the definition adequately addresses the broader interests of international and intercultural technical communication. The term localization has gained so much currency that virtually all translation services and their clients seem to use the terms translation and localization interchangeably. In fact, for good-quality translation, we might suppose there should be no difference; that is, a good translation is one that can be understood by the target audience. This means it must work within the cultural context of the audience—it "considers language, culture, customs, technical and other characteristics of the target locale."

But translation may have a more restricted meaning. Consider this definition from a 1965 monograph: "*Translation* may be defined as follows: *the replacement of textual material in one language (SL [source language]) by equivalent material in another language (TL [target language])*" [19, p. 20]. This definition concerns itself only with linguistic factors (although later in the book the author does recognize the need for at least some contextual, including cultural, accommodation in translation). This approach to translation is certainly used in some situations. When cost and haste are regarded to be of the essence, "equivalent" translation may be all that is possible. This is especially likely to be the case where machine translation may be considered "good enough." And this is the reality within which a number of our writers are caught: Their goal is to do the best translation that necessarily involves localization, but their budget, schedule, and work process determines that they often have to settle for "equivalent" translation while continuing to fight for good translation with localization. This internal conflict makes them think of the two concepts as existing on a same continuum. We decided that in our collection we will respect the current industry reality about translation and localization and not arbitrarily separate the two.

However, the field of localization and translation is also expanding and professionalizing as globalization of business becomes increasingly the norm. In the early 1990s, translation service for the corporate sector was primarily a cottage industry. By 2005 many established translation businesses were rapidly becoming consolidated into larger companies through mergers and acquisitions. "The combined turnover of the 15 biggest translation companies in the world represents 10% of the world market and 50% of the market for translation companies" [20, p. 24]. Meanwhile, startups continue to join the field; in that year (2005), there were 1,500 translation companies in Europe alone [20, p. 26].

This expanding market continues to attract newcomers and the competition may further heighten the expectations for quality. Along with such pressures comes a need for standards, which have both positive and negative consequences. Standards can be a positive factor in terms of helping to define what constitutes quality in localization. However, they can work in a contradictory way by imposing uniformity on practices that by definition and necessity should be responding to specific contexts and exigencies that standardization may not be able to accommodate. The localization industry is still very new and the needs they have to meet are often unprecedented. Training for professionals in the field, not surprisingly, is considered inadequate to meet the demand for practitioners. Thus, even as the localization profession grows, it is still struggling to define itself. Indeed, we predict that it has far to go as new areas of the developing world are brought into the global marketplace. We believe that a number of the stories in this collection provide a glimpse of the needs, challenges, and opportunities that localization practices bring to the work of technical communicators.

FINAL THOUGHTS

We tell this story of the stories not to show the trials and tribulations of editing this collection—it was neither—but to show our own learning process. It is not surprising, in retrospect, that telling stories based on personal experience is not likely to happen instantaneously and not without some sense of risk. Fortunately, our writers are not only experienced communicators but also untemperamental, resilient professionals. Although few of them have probably ever before written anything like the narratives in this book, they worked without complaint through revision after revision, completing the task in less than a year while working full time at their real jobs and—we hope—having a life of their own!

Now, even as the book is completed, we cannot say our learning process has ended, for we do not think we have everything figured out. What an interesting (but then again, inevitable) way to realize learning! We are grateful for having gone through this process, for the confusion, reflection, and negotiation that have happened to us along the way.

We hope that our readers will experience a learning process of their own when reading this book, and we hope that after using this book they will likewise appreciate the fact that they do not know it all and must therefore become life-long cultural learners. We hope each story, each interaction, each question, each answer, and each reflection will result in ongoing cultural transformation in everyone involved. We hope, as well, that they help readers get ever closer to understanding their own cultural identity, to appreciate another cultural perspective, and to be more skillful, critical, and ethical in negotiating cultural encounters.

REFERENCES

1. J. Allen, "The case against defining technical writing," *Journal of Business and Technical Communication,* vol. 4, no. 2, pp. 68–77, 1990.
2. R. J. Connors, "The rise of technical writing instruction in America," *Journal of Technical Writing and Communication,* vol. 12, no. 4, pp. 329–352, 1982.
3. C. R. Miller, "A humanistic rational for technical writing," *College English,* vol. 40, pp. 610–617, 1979.
4. A. Penrose and S. Katz, *Writing in the Sciences: Exploring Conventions of Scientific Discourse.* New York: Longman, 2010.
5. D. A. Jameson, "Telling the investment story: A narrative analysis of shareholder reports," *Journal of Business Communication,* vol. 37, no. 1, pp. 7–38, 2000.
6. J. Perkins and N. Blyler, "Introduction: Taking a narrative turn in professional communication," in *Narrative and Professional Communication,* J. Perkins and N. Blyler, Eds. Stamford, CT: Alex Publishing, 1999, pp. 1–34.

7. G. J. Savage and D. L. Sullivan, *Writing a Professional Life: Stories of Technical Communicators On and Off the Job*. Boston: Allyn & Bacon, 2001.

8. R. Johnson-Sheehan and A. Flood, "Genre, rhetorical interpretation, and the open case: Teaching the analytical report," *IEEE Transactions on Professional Communication*, vol. 42, no. 1, pp. 20–31, March 1999.

9. M. Markel, *Technical Communication*. Boston: Bedford/St. Martin's, 2010.

10. D. S. Bosley, *Global Contexts: Case Studies in International Technical Communication*. Boston: Allyn & Bacon, 2001.

11. M. A. Dyrud, "Book review: *Global Contexts: Case Studies in International Technical Communication*," *Business Communication Quarterly*, vol. 65, no. 4, pp. 127–130, 2002.

12. P. R. Hunsinger, "Culture and cultural identity in intercultural technical communication," *Technical Communication Quarterly*, vol. 15, no. 1, pp. 31–48, 2006.

13. A. Appadurai, *Modernity at Large: Cultural Dimensions of Globalization*. Minneapolis: University of Minnesota Press, 1996.

14. C. Geertz, *Works and Lives: The Anthropologist as Author*. Stanford, CA: Stanford University Press, 1988.

15. M. Foucault, "What is an author?" in *Language, Counter-Memory, Practice: Selected Essays and Interviews*, D. F. Bouchard, Ed. Ithaca, NY: Cornell University Press, 1977, pp. 113–138.

16. R. Kubota and A. Lehner, "Toward critical contrastive rhetoric," *Journal of Second Language Writing*, vol. 13, no. 1, pp. 7–27, 2004.

17. A. Wierzbicka, *English: Meaning and Culture*. New York: Oxford University Press, 2006.

18. Globalization and Localization Association (2011), *Localization Definitions*. Available: http://www.gala-global.org/view/terminology?filter0=&op0=&op1=starts&filter1= L&op2=contains&filter2, accessed December 29, 2011.

19. J. C. Catford, *A Linguistic Theory of Translation: An Essay in Applied Linguistics*. London: Oxford University Press, 1965.

20. A. Rinsche and N. Portera-Zanotti, *Study on the Size of the Language Industry in the EU*. Surrey, UK: Directorate-General for Translation of the European Commission, 2009.

CHAPTER 1

CHANGING TIMES, CHANGING STYLE GUIDES

Jennifer O'Neill

Jennifer O'Neill is a senior technical writer based in Brussels, Belgium. She has worked with technical publications for 17 years. O'Neill has a Master's degree in science in ergonomics and a background in usability. Prior to working with technical publications, she worked as an ergonomist in the United Kingdom and France evaluating the usability of buildings. It was during that work that O'Neill first became involved with building security and encountered the documentation that accompanies security products. O'Neill currently works for a U.S. multinational corporation that manufactures security products, such as closed-circuit televisions, for a global market. Canadian-born Irish, she speaks English as her mother tongue and is fluent in French. She has worked in three countries: the United Kingdom, France, and Belgium.

CHAPTER SYNOPSIS

Based on the writer's personal experience, this is a story about change, change that occurred against the backdrop of the global economy, change that rose out of the need for companies to stay fiercely competitive in the global market. As European and American companies went through mergers and acquisitions, publication units were relocated and restructured, and style guides had to be created, re-created, merged, and only to be abandoned. This story illustrates how a group of European technical communicators try to adapt to these changes and produce quality documentation for the global market. However, as various moves and false moves were made, the demands of some regions and markets

Negotiating Cultural Encounters. Edited by Han Yu and Gerald Savage

17

rose to the top, while the needs and concerns of others fell by the wayside. Debates and discontent broke out, and compromises as well as one-sided decisions were made. Through it all, we learn the challenges faced by European writers trying to produce, on tight budgets, documents for a regional market that operates in 20 languages. We understand their frustration trying to educate American writers and editors on how to write for translation and localization. And, probably most importantly, we are asked to accept the necessity and reality of change.

> The times they are a-changing.
> —Bob Dylan, 1964

In January 2007, three heating and ventilation manufacturing companies, two located in the United States and one in Europe, merged to create a global company. Their joint product base now included heating, ventilation and air conditioning (HVAC) units, boilers, pumps, fans, filters, refrigeration, solar panels as well as more software driven services such as energy management and building automation systems. The merger made sense as all three companies wanted to expand their global reach in an increasingly competitive marketplace. It would help jump-start long-term growth because each company brought business, technology, and market strengths to the table that complemented each other.

The two U.S. companies mainly operated in the North American market, although one had started to expand into Latin America and had small but growing sales offices in Mexico and Brazil. The European company operated across Europe, the Middle East and Africa (EMEA). From now on, all three companies would be known as Shannon Global Facilities, Inc. (all company and character names used are pseudonyms).

TECHNICAL WRITING TEAMS

The two U.S. companies both had a technical publications department, located, respectively, at their R&D sites in California and Arizona. Collectively they had nine writers, one editor, and two documentation managers.

The California writing team was based in Oakland, California. As a result of the merger, this R&D and administration site was now to become the global headquarters of Shannon Global Facilities, Inc. The writers were managed by Nancy Sherbakov. Her technical publications department had five writers and an editor. Nancy had a long career as a writer and editor, working in several sectors such as telecoms, financial and equipment manufacturing. She had been managing this department for the last five years. Zac Browning managed the Arizona technical publications department. He had built his team over the last seven years, growing it from just two writers attached to engineering to a technical publications department with four writers and himself as the manager.

The European company involved in the merger had four writers located in three countries—France, the Netherlands and Hungary—to accommodate the company's R&D sites in Maastricht, the Netherlands; Budapest, Hungary; and its headquarters in Lyon, France.

The European writer based in Lyon, France, was Sara Mitchell. She was English and had moved to Paris after graduating in French over 12 years ago. She had joined her current company six years ago from Paris as a senior technical writer. The move to Lyon had initially been a big social change for Sara, but now it was her home. It was the gastronomic capital of France and let her explore her love of cooking. It was also cheaper than Paris to visit restaurants renowned for their cuisine.

Sara's technical writing colleagues in Maastricht, the Netherlands, had also been moved around. Dirk was Dutch but was born in Surinam. His family returned to the Netherlands when he was nine years old. He lived over the border in Antwerp, Belgium, for a couple of years after graduating in journalism and then in Madrid, Spain, for a few years before moving back to the Netherlands. He had excellent English, was also fluent in German and Spanish and had reasonable French.

The other "Dutch" colleague was Sonia, who had been with the company for two years. Sonia was actually Scottish and had an IT background. She had a wanderlust spirit. So far, she had worked in five countries (England, Germany, Italy, France and now the Netherlands), usually as a contractor. Technical writing as a career gave her the flexibility to move around Europe for work. Sonia was fluent in German, had reasonable Dutch, and had some Italian.

Gabor at the Budapest, Hungary, office was Hungarian and had never lived outside his country. He was a young kid when the Wall fell in 1989 so he had learned English in school but had some knowledge of Russian. He had moved from technical support in the company to become a technical writer three years ago.

The European company had decided several years earlier that they would use U.S. English in all English language communications. So despite their diverse backgrounds, the European writers wrote in U.S. English. Their writings included installation, configuration, and inspection manuals for installers (which averaged around 80 pages) as well as a few end-user guides. The European writers also oversaw the localization of the manuals each had written, so they worked closely with the translation agencies and the sales offices across EMEA.

Now part of the EMEA division of Shannon Global Facilities, the European writers operated in 20 languages across the EMEA region. However, it was too expensive to translate the manuals into all these languages to meet customer needs, so languages were prioritized, depending on the product and its market. The average number of languages in a translation project was 12 (with an upward trend). They were paid for from the central translation budget managed from the Lyon headquarters. The demands of localization meant that the European writers were kept continually alert to the costs of getting their work to market in all the required languages.

PROBLEMS IN U.S. MANUALS

Shortly after the merger, Shannon Global Facilities started their efforts to sell in the EMEA market those products that were previously only available in North America.

In May 2007, Sara was told by the EMEA product management team that as part of her tasks she would now also be overseeing the localization of U.S. manuals and energy management and building automation system software for the EMEA market. They told her that the United States had previously only translated a few short manuals, which was done by the Mexican and Brazilian sales offices. The U.S. staff had not yet used the services of a translation agency as their sales offices translated for free. As Europe already had several years' experience working with many languages, the United States would now also send their work to them to be translated for products sold in EMEA. The EMEA business would pay the cost of these translations.

That same month, Sara started to receive the source files of manuals from the documentation managers in the United States. The manuals varied between 40 and 120 pages and all were done in Microsoft Word. As she began to check the U.S. manuals for potential localization issues, Sara quickly realized that there were three main problems: (1) the manuals lacked global branding; (2) they followed conflicting style guides; and (3) they had not been created with translation in mind.

Although Shannon Global Facilities had by now existed for over five months, the company still didn't have a single corporate branding identity. Marketing was working on it, but it probably wouldn't be finalized for at least a few more months. Sara needed to get these manuals out the door in multiple languages as soon as possible. However, she was faced with manuals that had been branded prior to the merger with multiple branding identities. The company names and logos were different from those used in EMEA and had different fonts, cover designs, and colors.

In North America, there were sales teams for each business group (HVAC, renewable energy, and energy management), and each group had its own branding. The U.S. technical writing teams expected Sara to keep their branding in the translated manuals until the new corporate branding was released. However, in EMEA, the sales teams in each country sold all products (as it wasn't practical to have three separate sales teams in each of the 30 countries), so one common branding identity was used for the whole region. The EMEA product management wanted Sara to convert the U.S. manuals to the EMEA branding until there was a common global one. They were adamant that they didn't want a kaleidoscope of brandings released in the EMEA market before the global branding appeared, particularly as some of the American brandings meant nothing to EMEA customers.

Looking at the source manuals from the two U.S. companies and comparing them with the European manuals, Sara also noticed that each group's manuals followed different style guides. For instance, there were many ways of saying the same thing and terminology differed. These inconsistencies, Sara thought to herself, would have an impact on cost. Their translation budgets were never large enough to meet demand and their managers were always chasing her and her colleagues to control cost. Her company usually paid a lower price for translation agencies to translate identical, or nearly identical, texts. But all these differences in writing meant lower reuse of content and higher translation costs. Ouch!

Another problem she faced was that the U.S. manuals had not been written for localization. There were no allowances for the impact of text expansion that so often comes with translation. The U.S. manuals frequently used forced page breaks to position paragraphs nicely on the page, which would affect the layout of translated documents. One documentation group often placed sections inside text boxes, which meant that when translated, the content would expand out of sight in the text box and the box size must be manually adjusted. The manuals did not contain metric measurements, numbers lacked a leading zero, the contact information was U.S.-only contact, and so on, all causing more work for her and the translation agency— and, again, more time and cost.

The biggest problem of all was about the graphics. Most of the graphics had embedded text.

A big nuisance, Sara thought to herself as she sat back in her chair. Sara needed these manuals in 10 languages. If all source graphics are sent to the translation agency, for every graphic, each language will be placed in a separate layer in the graphic file. To insert the translated text in a graphic could take around 10 minutes per language. That's 1 hour 40 minutes per graphic. The agency charges 30 euros ($44) an hour for such work—not including the actual translation, but that's minimal— so this single graphic will cost 50 euros ($73) to put into 10 languages. If the manual has 10 graphics requiring similar work, it could cost 16.7 hours and 500 euros ($731). This graphic-related work could easily account for 3 to 5 percent of the total translation cost of a manual and delay its release. Although 3 to 5 percent can seem a small share, do several manuals and these extra costs and time delay would accumulate. With tight translation budgets and deadlines, such extra costs and delays are unwelcome and need to be avoided.

To avoid these costs and delays, Sara had to modify these graphics first. Many graphics, Sara noticed, were simply jpeg files, which meant she had to go looking for the source graphics first. She then needed to add numbered callouts to the graphics and place the associated text under the graphics. Although the text could be directly placed in text boxes over or around the graphic in Word, translation agencies strongly disliked this practice. It ran the risk of the text being accidentally omitted in translation (translators don't work directly in the Word file but in a translation memory tool). The box could also move around on the page due to text expansion so it could need to be manually repositioned.

She sent an e-mail to the U.S. documentation managers explaining that she had to rework their manuals due to differences in branding and concerns with translation. Although the U.S. teams were annoyed that their branding had been changed, they admitted they didn't realize that products could be sold differently in regional markets outside of North America and that this could impact manuals. They also acknowledged they had never had to consider the issue of translation and the cost involved until now.

COMMON STYLE GUIDE AND TEMPLATES

In the teleconferences that followed between the U.S. and European writers over the next number of weeks, everyone agreed that it was important that all the manuals looked the same and were written consistently. There should be a common style guide and templates. This became possible in September when marketing finally released the new branding guidelines and corporate logos for everyone to use worldwide.

In November 2007, Sara and Sonia from the Dutch office went to the United States to meet their American colleagues and start the work on building the shared style guide and templates. It was time to put faces to names and voices.

The meeting lasted two days at the company headquarters and R&D center in Oakland, California. Nancy Sherbakov, the documentation manager, welcomed Sara and Sonia. Since the merger, Nancy had become the main person of contact between the European writers and their U.S. counterparts.

One of the first things Sara and Sonia noticed was that everyone in Oakland worked in cubicles, whereas the company's European offices were all open plan, usually based around teams. Chatting with their U.S. colleagues, Sara and Sonia admitted that they didn't fancy working in the isolation of a cubicle. However, their U.S. colleagues hated the idea of an open plan as they would consider it too distracting. "*Chacun son truc* [Each to his own]," shrugged Sara.

Then there was the language. It was the first time in years that Sara and Sonia had attended a meeting where everyone's mother-tongue was English. They were the only foreigners. All conversation was in English only. "Gosh," thought Sara, "it's a bit weird not hearing chat in other languages." Where she worked at the EMEA headquarters, there were perhaps up to seven or eight nationalities and you could hear four or five languages spoken between colleagues. In her department of five people, each of them was a different nationality (French, English, German, Danish and Swiss) and chatted in up to three languages (French, English and German). She'd never worked in a monolingual environment. None of her U.S. colleagues were fluent in a second language. "But then," she thought, "most British people aren't either." The "Anglo Saxons" are renowned for being monolingual.

Once everybody had introduced themselves at the meeting, Nancy updated everyone on the company restructuring, which had been quickly gathering pace. They were in the process of merging the two U.S. technical publications departments. Because of product consolidation, there would be two fewer U.S. writers. Two writers were relocating from Arizona so all U.S. writers would now be based at Oakland. Zac, the Arizona documentation manager, had been let go and Nancy was now the sole documentation manager. Nancy's editor had left the company to pursue interests elsewhere, so she had hired a new editor, Peter, who came from the telecoms sector. Altogether, there would now be seven writers in the United States plus Nancy as manager and Peter as editor (an overall change from 12 to 9 people). These changes didn't impact the four European writers, who would continue to report to their local managers.

THANKS FOR NOT TELLING US

Peter then took the stage to explain his documentation plans for the group. They were going to move to Adobe FrameMaker® and get away from Microsoft Word. He'd been evaluating the different document types his group wrote and had started working on several templates.

What? A new tool?

Sara and Sonia sat stunned. This was the first they had heard of a tool change.

They assumed that such a fundamental change would be discussed as a group beforehand. They asked why this decision hadn't been debated. Peter replied that Frame was a superior tool to Word and suited the company's needs better.

"But what about the needs of other regional markets besides North America? Were we considered? No one asked us anything about the potential issues of changing tools," said Sonia.

Peter replied that there were no problems with translation when using Frame. Sara agreed. Translation agencies preferred Frame over Word. However, that wasn't the issue (besides not having been told beforehand of such a fundamental decision). Sonia and Sara explained to the group that Europe used Word as they had to share the source Word files of manuals with their sales offices and large customers, the distributors. Although Europe spent around 280,000 euros (about $410,000) a year on translating technical documentation across all product groups, it wasn't enough to meet all needs. The sales offices in countries with smaller markets often had to translate manuals themselves as the only way to get their local languages. And they used Word. The writers also shared source files with distributors who rebranded and customized the files themselves, often in languages that no one knew in-house. They, too, used Word. With the tool change, who was going to get all the files from Frame into Word?

"Bottom line," Sonia said, "we need to be able to share source files with internal and external customers across multiple languages. Why weren't we involved in this decision to change tools?"

Peter seemed somewhat annoyed by this unexpected turn of events. He hadn't expected these European business requirements nor did he realize the size of their translation budget. He also wasn't happy that they were sharing source files with third parties. He also began to realize, however, that he had no say on the file sharing. It was a regional management decision. In his previous job, he hadn't needed to consider whether regional markets worldwide operated differently in the company—it wasn't an issue. This company apparently operated differently from his previous one. Reluctantly, Peter accepted that the European writers would stay with Word while the U.S. writers would move to FrameMaker. Because Sara localized U.S. manuals, she had to use FrameMaker as well. Peter now would have to do Word templates, which he wasn't thrilled about. Sonia and Sara were never given a reason by Peter or Nancy as to why they hadn't been involved in the decision-making process. For the U.S. team, writing and translation were apparently two distinct and separate entities that didn't touch. At least this meeting should change that outlook.

For the development of the style guide, everyone agreed that it would be a collaborative effort and that it would include guidelines on writing for translation and issues of internationalization and localization. It was agreed that Peter would write a table of contents for the proposed new guide and distribute it to the team who attended the meeting for review. Work would then start on it.

RETURN FLIGHT HOME

After checking in at the airport to fly home, Sonia and Sara slowly meandered to their departure gates. Sonia walked up to a coffee shop and looked at Sara, "I'm getting a coffee. Tea for you? Oops, I must remember to ask for hot tea or you'll get iced tea!"

"Hot tea is perfect and no milk," replied Sara. "I've now the French habit of drinking it without milk. Very un-British, I know."

Sonia ordered the hot drinks and looked around for a free table in the fast-food area. "Let's sit here," she said pointing to a table that overlooked the shopping and food area of the airport. They dumped their bags on the floor and sat down.

Sara spotted a newsstand in the shopping area. "I've still to buy magazines to bring back with me. English language magazines are so much cheaper here than Lyon and there's a fantastic choice. Got a stack of paperbacks in that bookstore near the hotel."

"Me too. The extra kilos didn't cause any problems checking in."

Relaxing back in her chair, Sara reflected on the events of the last couple of days. "In spite of the 'surprises,' we had an interesting gathering. Everyone's excited about the new global doc standards. Was taken aback over Frame, though. No discussion."

"I don't think the tool selection was a We decision. We effectively walked into a meeting with some fundamental decisions already taken and off the table. Nice of them to say nothing to us about it beforehand. Made us seem somewhat secondary to their focus."

"Yeah," sighed Sara. "Peter seemed to have already decided a chunk of stuff before we arrived. Was a bit surprised to say the least that he's already started working on Frame templates."

"Yep. Our market needs were somewhat incidental to Peter's plans. Europe is responsible for translation and carries the cost—a significant cost—yet he never asked us a thing beforehand."

Sara laughed. "Do we exist in their working life as we don't report to them? Seriously, now he knows more about our regional situation in Europe...Yet he was, well," Sara said, placing the coffee spoon to her lips as she considered her words, "more focused on U.S. needs. I'm still not sure what 'global' means to him. Our market is outside his scope of experience. But he should have got the picture over the last couple of days that we're a more complicated market with our 20 languages and 30 countries."

"Sure hope so. We have to be included. It shouldn't matter that we don't report to them in Oakland. We're all focused on documentation."

They both sat silently for some moments, contemplating and watching the life of a busy international airport go past.

"He certainly didn't seem very enamored by Word...," Sonia said to no one in particular. "I think it annoyed him that we must stay with Word in Europe. We have no choice. We must be able to share source files with our internal and external stakeholders. Our 'world' isn't just Tech Pubs." She gestured quotation marks around "world." "We work closely with Sales and Tech Support too as well as distributors across a lot of countries."

Sara contemplated what awaited her: "I foresee problems with many of the smaller countries. I don't yet know how to get them the Frame docs in Word so that they can translate locally themselves. My instinct says it'll be a lot of work, which I don't have time or resources to handle. How do you tell the Finnish or Czech sales offices or whoever that they can't have the source files to translate themselves in Word and I don't have the budget to do it centrally? And most of their customers don't speak English. They're in a bit of a Catch 22 situation. They're pushed hard by management to grow their markets but we can't provide them with translated manuals due to the lack of a budget. So they're stuck to do it locally until they reach that critical business volume where a translation budget is allocated by management and we can then take over the translations." She shook her head and sighed.

Sonia laughed, "Oh, the look of horror on Peter and Nancy's faces when they learned that we give source files to customers. But they'd no answer on how to customize docs for distributors when you don't know the language. And it's all blessed by Legal."

"A thought just crossed my mind."

"Sounds ominous."

Sara sat forward, "When Thierry, my boss, told me that I'd now be localizing the U.S. manuals, he mentioned that the Mexican and Brazilian sales offices were doing some local translations themselves but they had no translation budget. Peter and Nancy never mentioned Latin America. What's going to happen with them? My understanding is that the company also wants to expand into these emerging markets."

They looked at each other in silence.

"OK, work on a global style guide that incorporates international issues is now ready for takeoff," Sara said. "There are already some skid marks on the runway, it's true, but...."

"Houston we have lift off!" Sonia finished. They clinked their paper cups together.

BETRAYAL

During the two months that followed the meeting, Sara and Sonia heard nothing from Peter about the table of contents for the style guide. Nancy had said he was busy on it. When Sonia asked her how it could take an editor many weeks to do a table of contents, Nancy didn't elaborate. Odd.

After two months of silence, they and the rest of the editorial group received an e-mail from Peter—with an attachment. It was a 150-page style guide and several Frame templates!

The European technical writers were stunned. What happened to the agreements made at the U.S. meeting? They had been expecting a table of contents for discussion, not what seemed to be a completed style guide where any discussion would apparently just be tweaks. But what left the Europeans speechless were a couple of sentences he had written in the end of the Introduction: "International issues are not considered. They will be added at a later date."

European management had not spent thousands of euros sending two senior writers to the U.S. to explain the importance of writing for an international audience and the issue of localization in order to get this statement!

Sara and Sonia felt betrayed. After a meeting between Sara, Sonia, and the

EMEA engineering and product management leaders, Sara sent the U.S. editorial team an e-mail, a very long one, on behalf of the European team explaining why they could not accept this style guide.

> From: Sara Mitchell
> To: Peter Smith
> CC: Sonia Cummins, Thierry Thévenot, Dirk van de Woestijne, Nancy Sherbakov
> Sent: Thursday 22 January 2008 15:35
> Subject: Draft style guide—European feedback
>
> Peter,
>   Thanks for sending us the proposed style guide and templates. You and your team have put a lot of work into them.
>   We were surprised and disappointed to receive such a complete and long guide. At the meeting two months ago in Oakland, it was agreed as a group that you would initially send a table of contents to the editorial team for discussion. We never received such a document in Europe nor were told that you had started work on the guide with your team in the United States.
>   Adding to this discontent, we feel our concerns on translation and localization have not been properly addressed although during the meeting we discussed the problems we're currently having and proposed solutions and best practices. You all agreed that it was important to include them in the style guide. It was also acknowledged that although the U.S. team didn't have much experience in internationalization and localization, the European team did. By working together we would share knowledge and experiences. This agreement now seems to have been ignored.
>   Unfortunately the European team cannot accept the current draft style guide because of the lack of international perspectives. Some examples of the issues we have with it are
>
> *Focus on the U.S. market only*
>   The guide is written with the U.S. market only in mind. The brief meeting in the United States was for a global style guide. Yet the templates only list North American contact details. The other regions (EMEA and Latin America) do have their own contact details but they are absent. Customers based outside of North America are only provided with an international phone number in the United States to call.
>   In the U.S. manuals we have received to date for localization, only imperial measurements have been used. There is insufficient mention on the usage of metric in this style guide.
>
> *Incomplete terminology list locked in the style guide*
>   We need an extensive terminology list of the terms we use in our

documentation and software to help standardize the company terminology. The list needs to include descriptions of the terms as well as incorrect terms that should not be used. We must be able to share terminology. Placing it in the style guide limits access to just technical writers. This list must also be available to a wider audience such as engineering, marketing, sales offices and translators.

Too long and difficult to share

The guide is too long and reads more like a tutorial. We have a wider audience for this information than perhaps you may have in the United States. As explained, we share our source files with other groups and companies such as technical support, engineering, sales offices and distributors. Some of the information in the guide is useful to them but we feel that they are unlikely to read a 150-page guide. A shorter guide with separate stand-alone modules would suit us better. For example, we recently produced a short guide standardizing how to write metric measurements in technical specifications. It was well received in Europe and we'd like to see it included in the style guide.

Inadequate localization guidelines on handling graphics

A frequently encountered problem with manuals received from the United States is that of text associated with graphics. The guide says nothing about the problems related to embedding text in graphics, for example. We are not prepared to pay the extra translating charges incurred with embedded text as such costs are easy to prevent.

The guide also says little about the impact of text expansion that follows translation. FrameMaker allows writers to place text boxes on graphics in the document (so no need to embed text in the graphic) yet often writers are placing too much text around a graphic, which then swamps a graphic when it is translated. Callout lines are often placed too close together to accommodate translation. All these issues were discussed with you at the meeting. There must be guidelines on how to handle them.

Inefficient use of space in templates

The templates are excessive in page count. There's too much redundant white space such as heading levels starting by default on the next page and the excessive paragraph spacing around headings. The manuals of many of our products are printed in multiple languages, not just English, so printing costs matter to us. These excess pages increase the cost of printing, whether by us when shipping printed manuals with products or by our customers who receive PDFs and must print the manuals themselves. We need the layout to be tightened up.

Incorrect use of safety icons

The internationally recognized safety icons for Cautions, Warnings and Danger are incorrectly used in the templates. The IEC/ISO safety symbols have precise meanings. You've allocated different functions to them. This is totally unacceptable. You've used the Radioactive icon, for example, to identify all Danger notices in our manuals. None of our products is radioactive. The radioactive symbol must be removed immediately. You've also used the Electrical hazard symbol to identify all Warnings. Again this is incorrect. These safety symbols identify the type of hazard, not the level as you have done. The level of risk is indicated by the words "Caution," "Warning" and "Danger." Color can also be used to identify the level of risk. Our Legal department agrees with our concerns on this incorrect usage.

We need to discuss our issues with the guide so that they can be rectified and an international perspective integrated into it before the guide is formally released.

Cheers,
Sara
Senior Technical Writer

Over the next month several e-mails were exchanged between the U.S. and European members of the editorial team about Europe's concerns with the "global" style guide. Nancy held one teleconference during which Peter explained that in order to speed up the release of the style guide and templates, he and Nancy had decided to first release for the United States only and that "ROW [Rest of World] can do [your] own thing. Over time we'll merge."

So the meeting in Oakland where everyone agreed to a style guide that encompassed global issues was just a charade? The question about why they had never discussed this U.S.-only development with their European colleagues just drew silence.

REWORKING THE "GLOBAL" STYLE GUIDE

Sara and Sonia, with the support of their management, would not agree to an initial U.S.-only style guide. The approach didn't make sense/cents. The manuals written by the American writers were now being used internationally and translated. The European writers should be able to advise on international issues in the style guide. During the teleconference, Nancy agreed to set up a four-person editorial team to modify the style guide to include translation and localization issues. The team would include Peter and a senior writer from the United States, Richard, as well as Sara and Sonia from Europe.

Clearly there were expectation problems between the U.S. and European writing teams. The United States was keen to release the new style guide as soon as possible even if it applied to the United States only while Europe wanted this guide to also encompass international issues as everyone was now working for an international market and it was important to write the manuals for it to help control costs and improve time to market. The expertise to do such a guide existed in-house even if perhaps not in Oakland. A new style guide was a perfect opportunity to include international issues. The nine-hour time difference between Europe and California also hampered communication. Compounding this, Sara and Sonia tended to come into work earlier than most colleagues to avoid traffic congestion while Peter tended to arrive later at the office for the same reason. These habits simply widened the gap between them all. There would have to be another face-to-face meeting.

Sara and Sonia both felt that this second face-to-face meeting would be an excellent opportunity for Peter and Richard to come to Europe as neither had ever been outside of their country. Perhaps the fact that they had never traveled overseas might be part of the problem: "ROW" can seem too abstract and secondary from afar. Peter and Richard loved the idea of visiting Europe!

Unfortunately, the U.S. management said no to them traveling—no budget. So Sara and Sonia would go to the United States, again—European management considered this global project important. EMEA needed documents that were internationalized and easy and cost effective to translate.

Everyone was disappointed that the Americans could not come to Europe. Sonia thought it bizarre that Nancy, as documentation manager, now had "global" in her job title but had no passport or travel budget. "There's a potential credibility problem," she thought. In Europe most people have a passport. True, countries are much smaller in size than the United States so it's easy to cross borders. It is common for people living in the south of the Netherlands, for example, to go shopping in the large supermarkets in Germany. Prices are cheaper and it's the same currency.

In April, 2008, the editorial team met for three days in Oakland again to update the style guide. They had to put aside any feelings of mistrust and agree on what to include as Nancy had given them a rigid deadline: Official release of the "Global Style Guide" and templates would be the first of June, just six weeks away.

For Sara, her focus was on the biggest problem encountered: graphics. The U.S.-focused style guide had only a short chapter on handling graphics. It wasn't enough for such an important topic. When she had contacted the U.S. writers with whom she worked, they would invariably say they had no time to do what she requested or that they didn't agree with numbered callouts: They weren't user friendly. The cost impact wasn't their problem. There was also a cultural issue with using numbered callouts. European readers are probably much more used to seeing them in documents than U.S. readers. The problem then became clear: The U.S. writer is now writing for an international audience and the cost issues of going global cannot be ignored.

To push the graphics issue, Sara brought a draft of the proposed graphics chapter which included the information she needed and showed good and poor practice. Some of her suggestions were

1. Do not create or add callout lines to graphics in the documentation tool itself as they can move around on the page during translation and must be manually repositioned. Use graphics tools for graphics. Too often writers were doing shortcuts by drawing lines or boxes on graphics in Frame and Word.

2. Plan for text expansion when allocating space for text and positioning the callout lines.

3. Use numbered callouts instead of text callouts when there are several items or the items have many words to allow for text expansion in translation.

4. Do not embed text that needs to be translated (measurements, however, could be included in a graphic such as "1 in./25 mm" as they are rarely translated).

Sonia also brought the guidelines on writing metric measurements, planning to propose that they be included in the style guide. The guidelines covered such topics as capitalization, decimal places, dates, greater than and less than, multiplication sign, percent, range of values, weight, and SI units of measurement (Système International d'Unites, the metric system). Some examples from the guidelines were

1. Use SI units. Convert any measurements written in the Imperial/U.S. system to its metric equivalent. In documents created in the United States, use Imperial/U.S. measurements with the metric measurements in parentheses. Translators are not expected to do the conversion.

2. Don't use any decimal places when writing dimensions in millimeters. None of our products require this level of accuracy. Problems of excess accuracy usually arise from converting Imperial or U.S. measurements to metric.

3. Don't use the @ character as an abbreviation for "at" as it often cannot be translated.

4. Values less than 1 kg should be written in grams.

The two-day meeting went well. Being face to face certainly made discussion much easier and it was easy to show, even quickly draw up, particular points as questions came up. Using PowerPoint in a webinar didn't allow the same spontaneity. As they got to know each other better, it also became easier for the two sides (the United States and Europe) to make compromises without feeling that something important had been surrendered on bad terms. They also learned much about how the two regions worked and got a better understanding of why differences existed. A lot of ground was covered in those two days!

THE LAUNCH . . . AND THE UNEXPECTED

After the two-day meeting, Peter began working on implementing all the agreed changes to the style guide. He had a month to update the—now-global—style guide. Sara and Sonia would be returning to the United States for the launch of the global style guide in June, and it was agreed that both would give presentations to the group on how to write for the international market and how to do graphics for translation.

And then the bombshell hit. On May 6, 2008, Shannon Global Facilities Inc. announced that it had acquired one of their largest competitors, Reynolds Energy Inc., a manufacturer of HVAC systems as well as refrigeration, air treatment, dust collectors, filters and fans.

Nancy sent an e-mail to all the writing teams worldwide with the news, describing the newly acquired technical writing team. Reynolds Energy's technical writing department was based in Dallas, Texas. They had six technical writers, an editor and a documentation manager, all located on one site. Nancy confirmed that they would be sending their editor, Steve, and two writers to the official launch in Oakland in June to meet everyone and hear about the global project.

A global launch planned for one team was now being made for two. What would the new team make of it?

MAASTRICHT, JUNE 2008

Sonia dumped her backpack on the desk, took her coat off and shook the rain off it, "I'm back from the sunny, dry lands of California."

Dirk was sitting at his desk, located a few meters away from Sonia's. "Welcome back. How was the trip? You missed some really wet weather while you were away."

"It was a long flight back due to the stopover. I still feel some jet lag. It's amazing how much one can spend while waiting for hours in an airport. Did I really spend that much on coffee?"

"You can claim it back in expenses. How did the global release go?" Dirk asked eagerly.

"The guide was well received and no one said anything about the release date being delayed due to us 'cranky' Europeans always complaining about the original guide being unsuitable for our international needs."

"What did the Reynolds team think of the guide? What were they like?" asked Dirk.

"They were charming. They don't translate anything at present. They will probably do so at some stage."

Sonia had by now turned on her laptop and was downloading the latest e-mails. "We discussed the languages we do in EMEA and working with our translation vendor. It's all a new world to them. They don't develop too much software. For the moment they've no idea about translation and localization. They have more questions than answers."

"What was their impression of the global style guide? Will they use it? I'm still not convinced we need 150 pages that at times reads like a training course in technical writing but that's what we got." Dirk sounded somewhat frustrated.

Sonia shook her head, "Nope. They'll be sticking to their own style guide and Word templates. They use Word. So that means we've now two different sets of Word templates following different style guides as well as the Frame templates. But they've agreed in principle with having the same documentation standards world-wide across the company. Getting there will be another story if our experiences are anything to go by."

Dirk stood up from his chair, "Coffee? This one's free, from the Shannon coffee machine. So what's the next step?"

"I'll wait till you get back with that coffee. I'll unpack the docs I brought back."

A few minutes later Dirk was back. He placed an espresso on Sonia's desk, together with a couple of sugar cubes. Sitting back in his chair, he looked expectantly at Sonia, "Well?"

"So the editorial work on the global style guide now restarts to integrate the Shannon and Reynolds guides and templates. The goal is that the two style guides should more or less provide the same information. Let's see what happens with the templates as they are very different. There'll be a three-person editorial team this time: the editors from Shannon and Reynolds and then an international representative, which would be one of us in Europe. But I won't be taking part. It's just too time consuming on top of all deadlines I have to meet."

Dirk shook his head, "I have no interest helping out on the global style guide either. The editorial teleconferences are held biweekly at 18:00, but I must leave work at 17:00 to pick up my daughter from the crèche and so can't attend. A lot of the work is done outside of our work hours."

"Is Reynolds interested in international issues?" Dirk continued.

"Seem to be. The editor of the new group, Steve, was interested in what Sara and I had to say. During the Q and A of my presentation on writing internationalized documents he made an interesting comment," said Sonia. "I had given examples of issues we've had with docs not being written for translation and the impact it can subsequently have on the cost. He said that it was not the job of technical writers to get involved with cost issues. I was gob smacked!" Sonia made an "amazed" look at Dirk.

Dirk just laughed, "He said that? As soon as they start translating, they'll change. And their managers will insist in cost control too, just like ours. The English source document is where planning for localization starts. They'll get money conscious."

Sonia laughed too, "He doesn't work in a regional market with 20 languages and a tight budget. I told him that once his group starts localization, the cost issues will rapidly climb up his priority list."

"So what's next?"

"Sara and I now need to do a write up for our managers on the visit. The overall summary would be that we had a successful launch of the global style guide and now it restarts all over again due to the acquisition." Groan.

WORK STARTS, AGAIN, ON THE STYLE GUIDE

In September 2008, the new three-person editorial team started work on integrating the two groups' style guides. Peter represented the "global style guide" and Steve the "Reynolds Energy style guide." Gabor in Budapest was now the "international representative." He was delighted to be representing his colleagues in Europe.

They held biweekly teleconferences to discuss each section of Peter's style guide, compared them to the equivalent in Steve's guide and then tried to reach agreement on common guidelines. They issued minutes so everyone was kept updated and could provide feedback to their teams.

With 150 pages, there were many sections to consider. This was not going to be a short review process, Gabor realized. And he faced an unexpected complication. He was familiar with Peter's voice from occasional teleconferences that were held by Nancy with all the writers. But prior to these teleconferences, Gabor had not heard Steve speak, and Steve's accent came as a shock! Gabor found it strong and indistinct. It was initially difficult to understand everything that he said. Speaking on the phone further reduced clarity and he had no visual cues to help with the comprehension. What's worse, the economic situation had hardened and business travel was now frozen, so there was no money for Gabor to go to the United States and meet the others face to face.

Gabor didn't have the courage to tell Steve that he found his Texan accent difficult. He felt that if he said anything about the issue, Steve and Peter would question his English skills and his role as a writer. Although Gabor's English was good, he felt self-conscious, in the presence of two editors, that English wasn't his mother tongue. Gabor didn't say anything about this to his European colleagues either as he didn't want them thinking he might be letting them down.

So for the first few teleconferences Gabor didn't contribute much. He relied on Peter's input to try to follow the discussion. Gradually his ear grew accustomed to Steve's accent and his confidence and contribution to the discussion increased.

Slowly, bit by bit, the team tried to reach consensus (or make compromise) on the various items in the two style guides.

Waiting for her laptop to download some latest software updates, Sara read the minutes from the mid-February editorial meeting sent by Gabor. Progress seemed slow. The differences between the guides and templates still seemed distinct. It was hard to believe sometimes that both guides complied with the same corporate branding guidelines. They both used the approved fonts, although one guide opted for sans serif only in manuals and the other guide for sans serif for headings and serif for body text. Peter's guide used the company color in the headings and as shading (such as in tables). Steve's style guide was black with grey shading; no color. Peter's templates had pictures of the products on the front covers while there were no graphics on Steve's covers though both positioned the company logo in the same place on the cover. One wonders whether "branding" was being used by the editors to justify some of their "personal" style points in manuals.

As the alignment of the two guides seemed to be going slowly, the European writers began to discuss this issue in their monthly teleconferences.

March 6, 2009: Monthly Teleconference Between the European Writers

Sara: The editorial team has been working away for seven months now on aligning the style guides and, well, to be honest we still seem to have two distinct style guides.

Dirk: I know. Both editors say that each complies with the branding guidelines, but the guides as well as the templates look so different.

Sonia: The branding guidelines are really to keep the company look consistent and ensure we correctly use the approved fonts, colors as well as the company and product logos. If you like, to ensure a high-level consistent look. I don't think marketing intended to microcontrol technical publications.

Gabor: I asked at the last editorial meeting if it was the long-term policy to have two distinct style guides and no universal style guide. If it was, then we've wasted much of the last seven months going nowhere. Both Peter and Steve said that it was their goal to have a single global style guide. Steve added that it would take a long time getting there. I think it was Steve who said, "We can't simply 'switch over' to a single guide without risking serious disruptions."

Sara: Why not? We've done it before with Peter's style guide and templates.

Sonia: We're looking at the reality of style guides being a very territorial document. Let's just hope there isn't another acquisition and a third style guide walks in the door.

Gabor: I told Steve and Peter that they needed to get their managers to bring up this issue with the business unit boss to whom both tech pub departments report. The editorial meetings by themselves aren't doing it. At times it feels like I'm going to the dentist to have a blocked caption pulled. And two dentists are pulling it.

Sara: You poor thing. Well, all this style dental work is out of our hands. U.S. management will be the decision maker here. We're just spectators.... OK, let's see... next item on our monthly agenda is in-country reviews of translations. The translation agency wants to know how we can speed them up as some reviewers are taking way way too long.

MERGING OF TECH PUB DEPARTMENTS: BACK TO A SINGLE STYLE GUIDE AGAIN

In May 2009, after nine month's work, the updated style guides were released. Although the original plan had been to integrate the two style guides, they were still distinct guides. True, differences had been softened but.... The two editors still claimed their territories, even after all those months of work. This time there were no face-to-face group meetings to make the announcement. Business travel was still frozen due to the challenging economic situation. The release was all done by e-mail, not even a teleconference. There was no sense of drama or buzz like last time. Each writing group simply received a copy of its own updated style guide.

However, there wasn't much time to settle down with the latest guides before change happened yet again. Within weeks, Nancy left the company to pursue a career more focused on marketing communication. In July, barely three months after the joint release, U.S. management announced that the two separate technical publications departments at Shannon and Reynolds would now be merged, effective immediately.

The European writers would now report to the United States. In the future, there would be just one global style guide and everyone would use the same documentation tool, Microsoft Word. After barely 18 months, Adobe FrameMaker would be abandoned. No sooner had Peter's team got most manuals into Frame than they'd now have to move them to Word. Life is circular.

Peter and a contract writer from his team were let go. Steve's manager, Susan, was now managing 17 writers scattered across different sites and continents. Steve became the global editor-in-chief. And Peter's style guide was dead.

The suddenness of the change caught everyone by surprise. Sara felt disheartened at the loss of all the work that had been put into the Frame-based style guide and templates. However, Steve did say that over time he would move many of the useful items from Peter's guidelines into the new global style guide. He told Sara that eventually he would include the information on doing graphics, together with examples, in the style guide but he would not be using the ISO safety icons. He considered such items "visual clutter."

It was frustrating that after nine months of "consolidation," many of the graphic recommendations for localization still weren't in Steve's style guide. Sara didn't know how he was going to include the examples as his guide was a Help file with no graphics.

"You reach a stage sometimes," Sara thought, "where you just don't have the energy to comment."

But another part of her was optimistic. Now she would again be able to share the source Word files with her internal and external customers. This was good news as the sales offices had been getting increasingly discontented and vocal about not being able to translate manuals themselves and had been complaining to upper management in Lyon. Her management knew she couldn't do much to help them; she simply had no time to convert the Frame files into Word and there was no budget for a translation agency to do the work. Now the documentation could again become more "customer" friendly and not just focused on being end-user friendly.

U.S. management had put Steve in charge of creating international technical documentation. Steve was now responsible for guiding the company's English and translated documentation and software/firmware. He was genuinely keen to learn more about translation, localization and the documentation needs of the different regional markets. He admitted that he had little or no experience in these areas. So yet again the European writers had a person in charge who knew little about these important areas.

How was someone without experience in these areas supposed to "guide" such processes in a timely manner and on budget? Would they ever get someone who actually had trench experience of international work before becoming responsible for global documentation? *Plus ça change, plus c'est la même chose,* as the French say.

Mergers and acquisitions (M&As) invariably entail company restructuring and changes to business needs. They can also bring together disparate groups often spread out geographically with different cultures and experiences. A successful merger of teams needs open communication and a clear explanation of the new goals. Everybody needs to feel part of the new structure and that their skills are valued. Trust matters and it is something easy to destroy and difficult to rebuild.

The team leaders need to be honest. If they lack in-depth experience in areas that are now needed by the company, such as in internationalization and localization, and others in the merged team do have such experience, then the leaders should not be afraid to seek the active participation of those with that expertise, even if the expertise is based half a planet away and not next door. Control cannot be narrowly focused on a small group with little international experience. If a team feels that its culture, skills, and regional market needs have not been recognized by the "ruling" team, this can quickly create tension. The "subordinate" team may decide to "fight back" for the recognition of their regional needs. Their local managers will often agree with this tactic as they too may feel that their business needs are not being met.

Furthermore, with repeated departmental restructuring that can happen with M&As, reshuffled teams may lose the gains made for their regional markets and must now restart their case again under a new leader. So it's important that leaders are aware of international documentation needs before making changes. Otherwise, working life really can be frustratingly circular in multinationals. Regional markets are often different. It is important that we have an understanding of what these differences are and how they can impact documentation. And an obvious place to start learning is from our own team members located around the world.

SUMMARY OF LIFE CYCLE OF THE STYLE GUIDES

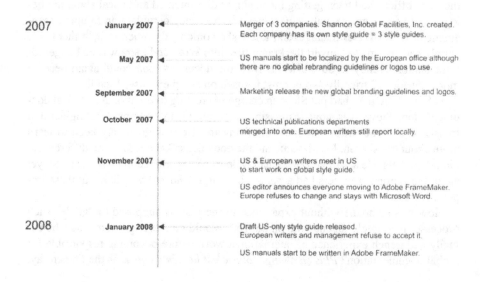

2007	January 2007	Merger of 3 companies. Shannon Global Facilities, Inc. created. Each company has its own style guide = 3 style guides.
	May 2007	US manuals start to be localized by the European office although there are no global rebranding guidelines or logos to use.
	September 2007	Marketing release the new global branding guidelines and logos.
	October 2007	US technical publications departments merged into one. European writers still report locally.
	November 2007	US & European writers meet in US to start work on global style guide.
		US editor announces everyone moving to Adobe FrameMaker. Europe refuses to change and stays with Microsoft Word.
2008	January 2008	Draft US-only style guide released. European writers and management refuse to accept it.
		US manuals start to be written in Adobe FrameMaker.

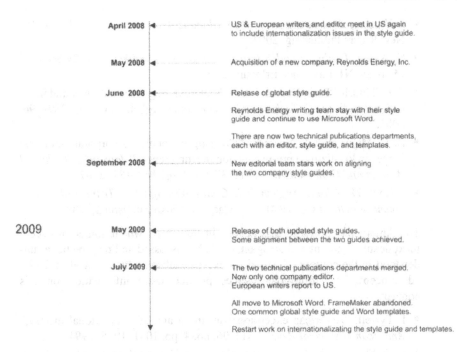

April 2008 — US & European writers and editor meet in US again to include internationalization issues in the style guide.

May 2008 — Acquisition of a new company, Reynolds Energy, Inc.

June 2008 — Release of global style guide.

Reynolds Energy writing team stay with their style guide and continue to use Microsoft Word.

There are now two technical publications departments, each with an editor, style guide, and templates.

September 2008 — New editorial team stars work on aligning the two company style guides.

2009

May 2009 — Release of both updated style guides. Some alignment between the two guides achieved.

July 2009 — The two technical publications departments merged. Now only one company editor. European writers report to US.

All move to Microsoft Word. FrameMaker abandoned. One common global style guide and Word templates.

Restart work on internationalizating the style guide and templates.

RECOMMENDED READINGS

1. This story brought up a number of concepts: translation, localization, and internationalization. These terms tend to be defined somewhat differently across sources and industries. The following readings will get you started in understanding them:

   - R. Ishida and S. K. Miller. (2010, Sept. 27), *Localization vs. Internationalization*. Available: http://www.w3.org/International/questions/qa-i18n, accessed April 23, 2012.
   - N. Kano and Microsoft (2012), "Chapter 1: Understanding internationalization," in *Developing International Software*. Available: http://msdn.microsoft.com/en-us/library/cc194758.aspx, accessed April 23, 2012.
   - J. B. Strother, "Localization vs. internationalization: E-learning programs for the aviation industry," in *Proceedings of E-Learn World Conference on E-Learning in Corporate, Government, Healthcare, & Higher Education, Montreal, Quebec, Canada, 2002*, pp. 895–900.

2. The story provides a number of techniques for translation and localization. To learn more about these techniques and the broader strategies of creating global and international documentation, refer to the following sources:

   - B. Esselink, *A Practical Guide to Localization*. Philadelphia: John Benjamins, 2000.

- J. Yunker, *Beyond Borders: Web Globalization Strategies.* Indianapolis: New Riders Publishing, 2003.
- N. Aykin, *Usability and Internationalization of Information Technology.* Mahwah, NJ: Lawrence Erlbaum, 2005.
- E. A. Thrush, "Plain English? A study of plain English vocabulary and international audiences," *Technical Communication,* vol. 48, no. 3, pp. 289–296, 2001.
- D. L. Major and A. Yoshida, "Crossing national and corporate cultures: Stages in localizing a pre-production meeting report," *Journal of Technical Writing and Communication,* vol. 37, no. 2, pp. 167–181, 2007.
- J. Kohl, *The Global English Style Guide: Writing Clear Translatable Documentation for a Global Market.* Cary, NC: SAS Publishing, 2008.

3. The Americans' attitudes and actions in this story, which sometimes seem inexplicably ignorant and arrogant, could be discussed and may be better understood in the context of American exceptionalism, which is itself a debated concept. The following sources provide more information on this concept:

- I. Tyrrell, "American exceptionalism in an age of international history," *American Historical Review,* vol. 96, no. 4, pp. 1031–1055, 1991.
- B. E. Shafer, Ed., *Is America Different? A New Look at American Exceptionalism.* New York: Oxford University Press, 1991.
- D. L. Madsen, *American Exceptionalism.* Edinburgh: Edinburgh University Press, 1998.

DISCUSSION QUESTIONS

1. Why is it important for technical writers and those who manage translations to work closely together? What feedback from the translation process could be useful to writers for their work?
2. How can writers increase their international experience if they are unable to travel for their work or do not directly work with colleagues in other countries?
3. The story discusses the work involved in translating and localizing graphics. Find some graphic examples from manuals that will be good as well as poor candidates for translation and localization.
4. What is the importance of using a style guide in technical communication projects, especially in a transnational company the size and magnitude of Shannon Global Facilities? And why would a style guide become "a very territorial document"?
5. Can there truly be a "global" style guide? What content should such a guide try to encompass?
6. Was it or was it not "the job of technical writers to get involved with cost is-

sues"? How might the answer to this question change when we consider both the domestic and international market? What is the implication of this question to technical communication education?

7. What impact can mergers and acquisitions have on a technical publications department and on writers? How can publication managers and their managers foster a cooperative spirit between writing teams pulled together by organizational change?

8. In the story, the European technical writers, because of their concerns and challenges, were frustrated by the American writers, editors, and management. What do you think were the American team's primary concerns? What might be the challenges that they faced?

CHAPTER 2

CONCINNITY OF ASSORTED NUTS

Marina Lin

Marina Lin is an information architect in the Chicago area. She holds a Master's degree in information architecture from Illinois Institute of Technology and currently works as an interaction designer for Mobile Apps at Cars.com. She also teaches and designs online courses in web design and development. She is particularly interested in user behavior when designing the best user experience at all interaction touch-points. Lin previously worked in higher education and e-commerce industries and specializes in website personalization as well as the intersection of design and business requirements. Lin is currently focusing on expanding her research techniques to include contextual inquiry, remote usability testing, and measuring user experience via key indicators. Her work has previously appeared in Intercom *and* Business Communication Quarterly. *Her full portfolio can be found at www.marinalin.com.*

CHAPTER SYNOPSIS

This story documents the experience of Marina, a user experience architect based in Chicago, as she worked with a globally dispersed and diverse team on an e-commerce project. The team was to build a web-based seller portal for a large retailer in the United States, helping the company participate in the booming e-commerce business. With a tight schedule and insufficient labor, the project setup was not ideal, but Marina was optimistic in making a difference. Quickly, however, Marina found herself on the defense side fighting for the end users while business expediency took the front seat. Although this may not be news to some usability experts, a number of other factors made the situation particularly trying for Marina. Multiple time zones and the inability to meet team members face to face had created communication barriers between them. Although Marina tried again and again to explain her work process and to invite collaboration, the conference calls were awkward and unproductive, and instant messages seemed all but a hit-and-miss game. Add to these the multiple and different personalities and communication styles, Marina almost wished she were never put on the project. Somehow, though, the project plunged on, not without some unexpected twists of events and interesting revelations for Marina.

Starting a new project . . . pretty excited.

My role as a user experience architect at a large national retailer headquartered in Chicago was about to get a lot more interesting, kudos to the growth of e-commerce and the latest trend of social networking which was spreading like fire in the online world. The idea was that if people are networking with each other online, they may want to consult their networks when making a purchase and essentially shop together online. With websites such as Amazon.com taking a lion's share of the e-commerce marketplace, our company was in a struggle to come up with online offerings unique enough to compete. Gone were the days when user experience could be sub-par for a website to successfully sell products.

Half of the company's online development group was tasked with developing new interactive ideas to engage our customers, such as allowing them to share their shopping experience with their online network and tailor their browsing experience to meet individual shoppers' needs. This online development group consisted mainly of three departments that worked heavily with each other to launch each new project. The product management department represented the business needs, the user experience department represented the user needs, and the IT department made the website come to life. Each department had its own manager, that is, the product manager, the user experience manager, and the IT manager. Each project was, then, assigned a project manager from the project management department who was responsible for coordinating efforts in order to deliver the project on time. My own department—user experience—where I worked as a user experience architect was broken down further into several roles: user experience architects, web designers, copywriters, and front-end developers.

This large group of folks had been working together and growing in size for quite a while and had a fairly well-defined process. The process starts with the product management department outlining the business needs and requirements, which are passed to the user experience department. There the user experience architect goes through a discovery phase and a competitive assessment based on these business needs and requirements. Using the findings, the user experience architect then creates wireframes, which are similar to the blueprints a traditional architect would employ—only the wireframes represent web pages, not physical buildings. These wireframes detail the page flow, the elements on the page and their hierarchy, and document interaction. The web designer takes the wireframes and uses them as a guide to create actual page designs. The front-end developer then creates basic HTML pages and the baton is passed to the IT developer who codes the logic of the website. The pages are then handed off for testing to the quality assurance (QA) engineer, who is part of the QA branch of the IT department. Of course, there are also copywriters who write marketing and advertising copy for banners and landing pages and craft web content in straightforward and easy-to-understand fashions to help the audience wade through the website. Additionally, there are business analysts who are another branch of the IT department and who oversee the tie-in between the user interface and the back-end system. Presiding over this organized chaos is the project manager, who ensures the dependencies are met in a timely manner.

While the company was abuzz trying to capitalize on social networking, I, personally, felt it was not an immediately applicable idea, although it was certainly interesting. After all, online shopping was still a solitary pursuit and the customers I have interviewed during usability studies, especially customers of our company's traditional brand, showed little interest in sharing their purchases with anyone online. So, I didn't mind at all when I was put *not* on the social commerce project but on something different, something I felt could bring *real* value.

This new project I was assigned presented a way for our company to compete with Amazon.com in a more direct way—by establishing a true marketplace on our site that allowed other sellers to sell directly. This endeavor made much more sense to me in terms of business and customer value. Why not allow other sellers to create stores on our website and sell their own products using our logistics, operations, and website framework? The company revenue would primarily be derived by charging commission on all items sold through our site and possibly additional fees to join the program. With other less tangible concepts floating about, I was thrilled to be placed on a team that would create a portal for these sellers to set up their stores from scratch. We were given a launch date of three months from the start of the project, and I was assigned the role of the lead user experience architect.

Because the company had put forth heavy efforts on the social commerce innovations, we had narrow internal bandwidth to fill the needed designer and development roles. Since the project launch time could not be pushed back, it was time to look externally for help to get it done on time. Our company's California office had a small branch of developers who happened to be of Russian heritage. That group typically worked on a different website that was part of the larger corporate brand so I had never crossed paths with them. Given the tight schedule, this was as good a time as ever to bring them onboard. Because the California office didn't have onsite QA staff, these developers had outsourced their quality assurance work to engineers from Belarus, most likely due to personal connections since Belarus is a former Soviet Union republic. Also brought into our seller portal project was an external design agency based in Bogota, Colombia, which we sometimes commissioned for project overfill.

The rest of the team was local, from the Chicago corporate headquarters. Sandip was the product manager who set the business requirements for the project (all character names used are pseudonyms). He was fairly new to the company, but joining us from Amazon made him an authority on the subject of online marketplace. Laura, the project manager, and a copywriter named Keith were also assigned to the team, and I had worked with both of them extensively.

KICKOFF MEETING

In a professional environment, one comes to expect a certain behavior from one's colleagues. Regardless of their different roles or backgrounds, there seems a certain layer of business-speak that masks people's true personalities. For example, meetings often take place with a polite and noncontroversial tone, and people use and expect set phrases (i.e., "Let's take this offline," or "Actionable items," or even "What's the ask on this project?" which isn't grammatically correct but yet became a standard phrase as a synonym for "request"!). In fact, business-speak has evolved into its own lexicon and has online dictionaries available for quick translation. This particular way of speaking helps to homogenize interoffice communication and adds a touch of predictability to most projects.

This predictability was certainly not part of the seller portal project, right from the beginning.

Sure, the project began as it always does with a kickoff meeting, which was on a Monday morning. The team members, however, were not in the same room warily assessing each other. Instead, I sat alone with Laura and Keith in a conference room. Sandip, who worked out of a different office in Chicago, was on the phone. The Russian developers dialed in from California. The designers spoke up a few minutes late from Bogota. And the quality assurance team? I never even heard their voices. Not being remotely in the same time zone, I only saw lengthy e-mails from them as the website began to take an unwieldy life of its own.

We introduced ourselves to each other, but nobody was overly friendly or chatty. The first meeting was very short and only the deadline of the first deliverable and the fact that we had an immovable launch were discussed. We knew that we had three months to complete the project and we agreed that the first deliverable would be done by the end of that work week. The team members talked briefly about the process but glossed over many of the details (that I thought!) we were accustomed to from previous projects. My role was to cohesively approach the user experience and design the website to meet each business goal and objective. This meant I would be the central point of contact for every other team member. The first deliverable of coming up with the site architecture and the homepage layout was also officially mine. Laura asked if there were any questions and was greeted by a few mumbles and then silence. And so we were off.

FIRST PRESENTATION

As I started working on wireframes, my fellow user experience architects who were assigned to the social commerce projects plunged into their work too. Secretly smug that my project was so much more practical and valuable to the company, I lent a sympathetic ear as they complained about the chaos and lack of process on their project.

My own work was coming along nicely. I started with the sitemap and the landing page, which didn't have much functionality aside from the ability to begin the account creation process. The deadline of one week was pretty tight and I could not afford much time for the discovery phase. Ideally, I would have talked to potential sellers, observed how they did their work, and drawn conclusions on which features of the seller website they would find necessary. Of course, I would have collaborated with the rest of the team as every role, from the designer to the developer, has its own unique perspective and can add important details to the interaction design. However, since time was a major constraint, I decided to include this collaborative piece in our first wireframe review.

Not being in the same physical location, this turned out more challenging than I thought. On a hectic Friday afternoon, Laura, Keith, and I met in one of the many conference rooms in our office. We had to wait for others to finish their meeting and leave the room before piling in to set up the conference call. Laura, being meticulously punctual, quickly dialed the conference number and apologized to the Russian developers who were already on the line, waiting. She signaled for me to quickly set up the web meeting, a software program that allowed me to share my screen remotely with the other attendees. As I fumbled with my laptop, Laura and Keith seated themselves around the table and prepared to take notes. The Colombian designers joined the call while I was setting up. They said a quick hello and everyone went back to silence. Keith raised his eyebrows in encouragement as if sensing this was going to be a tough crowd. I finally connected my laptop to the TV in the conference room, projected my wireframes on the screen, and shared the view of the screen with the remote teams who would follow the wireframes on their own computers.

I explained that I had worked on the sitemap and the landing page as a start because, having very loose guidelines in place, I felt this should be a launching point for all of us to get on the same page. Laura and Keith nodded, but I was greeted by silence from the phone line. What was already an awkward meeting started to become more uncomfortable. Perhaps we should have discussed the process earlier and in more detail, or gone through what a wireframe is, or even shared a few personal stories—anything to break the ice. It was too late, though, because as these thoughts were going through my head, I was deep into talking (to myself, it seemed) through my design decisions and plans for the rest of the website. When I was finished, the reaction was this:

Colombian designers: "Sounds good."
Russian developers: "When do we get the rest of the site?"

I planned to explain why it was important for all of us to be on the same page, but the phone made a few brief clicks. Team members on the other end started hanging up as we ran out of time.

SECOND TRY

All right, so no collaboration and no discovery. Also, Laura started wondering why I was taking so long with getting the second set of wireframes for the account creation and login pages. I decided to continue with a quick competitive analysis by looking at similar websites from our competitors. Remembering the sage advice of never coming to a presentation without having at least one person solidly on my side, I also decided to have a chat with Sandip, the product manager, to get his opinion and buy-in prior to presenting the wireframes to the larger team. I could design the most elaborate interaction, which is relatively easy to do conceptually using prototyping software, but I needed to know whether it would work before I wasted time going down the wrong path.

It was tough to get on Sandip's schedule. He was busy planning the seller portal release from the business side and his attention was constantly demanded elsewhere. He also worked out of a different location in Chicago and mainly called into meetings along with other team members. We were finally able to coordinate some time out of his busy schedule when he was visiting my location early the following week. I reserved a conference room and ran into him as I was heading to the meeting. We said hello but he seemed a bit distracted and I felt like I was getting sacred face-to-face time with the president. I talked fast and showed him my wireframes. I rushed through partly because I knew our time was limited and also because I was nervous. Since Sandip wasn't in the first presentation meeting and I hadn't gotten any feedback from the team during that meeting, I wasn't confident at all I was on the right track. He listened patiently and nodded.

"Have you taken a look at Amazon?" he asked when I was finished.

I had, of course, been in the process of doing a competitive analysis of sites that offer similar capabilities, but this wasn't what he was getting at.

"I strongly recommend you familiarize yourself with their seller sign-up process and site architecture. We're going to be building something very similar."

"But I'm trying to figure out a way to provide a differentiator. We don't have to do it that way exactly, do we?" I interjected.

"Why not?" he seemed surprised. "We don't have much time and their way works, doesn't it?"

·I hadn't realized at the time that by "something very similar," his expectation was that our architecture would be identical. I did get the hint and was hoping to clarify, but his Blackberry piped up and he was off to his next meeting.

While I huffed at my desk, frustrated for not having an ally, the social commerce team was excited. They had just launched Pebble, which was basically Twitter for the corporate world. Pebble allowed the team to share work-related information, essentially immersing them in the same online social environment they were trying to create for their customers. Just like at Twitter and other similar sites, people posted ridiculously peppy nonsense that was so gung-ho it could make one gag. I thus proceeded to waste the hour following my discouraging chat with Sandip wading through other colleagues' exclamation-heavy posts and considering what my posts would look like had I signed up.

"Project going nowhere—can't wait for the next presentation!"

That second presentation went even worse. This time I had significantly more completed wireframes for review. Working off the product requirements document and, yes, copying aspects of Amazon, I started to document exactly how the account setup process will flow, which features it will include, and how the sellers will interact with the pages. As the presentation devolved from me talking and others doing who-knows-what on the silent phone to me growing more and more defensive, I realized we were anything but on the same page—as it turned out, both figuratively and literally.

"Why does clicking on this link take you to a different page?" inquired Yury, the Russian developer.

"Well, it's not a different page. It's a different view of the same page! See, the wireframes are set up to show conditions of the same page depending on ...," I started to say.

"I don't think Amazon does this," Sandip interrupted.

This presentation was getting away from me.

"This third page is very complicated and is totally out of scope for this phase," Vlad, another developer, declared, adding to what was starting to feel like a barrage of questions I wasn't getting a chance to answer.

"Fine," I sighed, "but why didn't you tell me this last time when I asked before I spent time working on it?"

"That's not how we understood it," Yury's voice conveyed a shrug.

"Ok, what do you think of a slightly different approach? We could combine several steps into one and then the user can accomplish this sign-up task in fewer clicks," I asked and was greeted by silence.

"What do you guys think?" I tried again.

There was more silence for a while and I imagined the developers exchanging looks.

"That's your job to figure out," they finally said and the meeting ended.

It was time for a time-out.

Laura, the project manager, informed me that due to this rework of the previous sections, I was now falling behind on starting the next batch of sections. I started to get really frustrated because this was exactly what I was consciously trying to avoid. Another frustrating thought also came up. I realized that not knowing how to properly read a wireframe was causing confusion during our calls. The wireframe format was familiar to the company's local IT teams but apparently not to the external ones I was working with on this project. The developers questioned why there were so many pages, but the designers never said a word. Since they, too, were new to this format, I assumed they might have questions too, but they remained characteristically silent.

So I e-mailed a brief tutorial to the entire project team, including Sandip, to explain how to read a wireframe. I explained that wireframes are different from mockups. A wireframe is primarily used to establish content and information hierarchy. It also establishes interaction rules and addresses what happens when users follow certain paths or encounter errors. It is very bare-bones with minimal graphics and color but instead contains a lot of annotations and flowcharts. The mockups, on the other hand, take the wireframe and dress it up to add form to the function. They show exactly how the page would look to the audience once it goes live with high levels of design detail. It was important for the team members to understand these differences and that, at this point in the process, we were still looking at wireframes to build consensus on the site flow and interaction. My e-mail received only one response, from Carlos, the lead designer from Colombia.

"Thanks!" he wrote, "this looks great!"

I hoped that the others' silence indicated that they read and understood what a wireframe was so that in the next meeting we could really focus on the wireframed concepts and the designers could get started on the mockups. This wireframe tutorial should be right in time, I thought, because in my next presentation meeting I had planned to present a recommendation for sellers to upload their merchandise.

THIRD TRY

After another week of catching up with the revisions of the first set of wireframes, Laura finally reserved time in one of the conference rooms early Monday morning for me to present the next batch of the seller portal pages, namely the upload merchandise section. Something happened at this next meeting that I had never seen before or since.

My concept was still very rough because I was hoping to get the team members to suggest improvements and brainstorm together. The deadline was tight, so my interface draft was also simple. I modeled some of the interaction on Amazon's interface, but the rest was unique to us. The sage advice of having someone on my side in the room rang in my ears as I headed to the meeting, and I had a sinking realization that, aside from Sandip's comments earlier and maybe Laura's and Keith's support, I wasn't at all sure how this solution would be received. I was optimistic, though, because I was hoping this would spring into a lively discussion and brainstorm.

For this particular meeting, Laura, Keith, and I were physically present. Sandip dialed in from a different location, and so did the developers and designers. As the typically awkward phone conference started, I pulled up the new documents on the shared screen.

"Why are there so many pages again? I don't understand...." Sandip's voice trailed off and I sensed mild annoyance.

Clearly, my tutorial e-mail had gone unread.

"Sandip, it's multiple views of the same page. You see, when the seller clicks on the link, this section expands but he remains on the same page. It's just that my wireframe shows this interaction in several pages so that I can...."

"Yeah, yeah," I was cut off by Yury as I started to launch into a recital of the tutorial e-mail. "Listen, we're going to put the phone on mute for a bit but you should continue your presentation and don't mind us. We'll be back in bit."

The line went silent. Laura, Keith, and I exchanged glances and I meekly asked the designers if they were still there. They cheerily replied that they were and that I should continue with the presentation.

And so, for the rest of the hour, we literally had two parallel meetings. I talked to a silent line with the developers, the designers, and Sandip saying nothing. I presumed the developers conferred on their own solution, which they proceeded to unveil after they took the phone off mute in the last five minutes of the meeting. They laid out their plan with a direct "Here's how we're going to do this" approach. Instead of my proposed interface, which provided a simple graphical interface for sellers to select products they wanted to upload either individually or in bulk, they stated that having the sellers upload an entire spreadsheet and confirm their upload, a two-step process, was the best way to achieve this task in the shortest development time. Uploading the entire spreadsheet was simpler, yes, but it left many ambiguities on whether the entire contents of the file were read correctly and there was no immediate way for sellers to know whether their upload was successful. But, this was a no-nonsense proposal that left little room for debate.

"Marina, this is what we had discussed, remember?" asked Sandip. I felt a gentle reproach in his tone and then remembered that Amazon treated this upload process quite similarly to what the developers had suggested.

"I'll update the wireframes?" This was more of a question from me because frankly I wasn't sure what my role was anymore.

"OK," Laura informed me, "but can you do this today? The designers have to get started as soon as possible." She stated this matter of factly because, although she was interested in the best user experience possible, it was also her job to keep this project humming along.

After the meeting ended, I told her that this was only supposed to be an initial concept and that I thought the rest of the team would provide their input instead of disregard it completely and do it their own way.

"They did provide their input, though, and now you have clear direction," Laura pointed out.

She was right, but I felt cornered and had all but given up.

New status: *"Went from lead to being marginalized. Isn't it great to feel so useful?"*

SKYPE WITH CARLOS

In the meanwhile, we were about three weeks into the project, and the user experience and design time was running out. The developers had to have as much time coding as possible and so the designers from Colombia began sending their initial designs based on my wireframes. Or rather, I assumed they were initial because they looked exactly, pixel for pixel, like my wireframes. Like I said earlier, wireframes are meant to establish hierarchy of content, user flow, interaction, and other documented elements. What they definitely are not is a final design, which was exactly what the Colombian designers took them to mean.

Naturally, I wanted to critique this design choice, but this was out of my jurisdiction as a user experience architect. We didn't have formal design reviews scheduled and Laura suggested that, instead of taking up more time with constant meetings, I should provide feedback to the lead designer over Skype instant messenger. Unlike our utilitarian working relationship with the developers, who never reached out outside our conference calls, the lead designer, Carlos, had introduced himself via e-mail prior to the project kickoff and sent me a Skype invitation.

After a brief, friendly back and forth about the weather differences in Bogota and Chicago, I asked whether the composite screenshots they had sent over had been finalized. I knew they had been, but instead of berating him, I decided to approach this in an indirect manner.

> Carlos: Yep! We're already working on the second batch and will have them to you shortly. ☺
>
> Marina: That's great, thank you! I was wondering if you might have any suggestions on improving the layout? My wireframes were meant to suggest information hierarchy and not so much the exact design.

There was a slight pause in the typing and I wondered if Carlos was conferring with the other designers on the project.

> Carlos: The wireframes look great, though. You did a really nice job. ☺
>
> Marina: Oh thank you, I appreciate that—Um, let me know if you have any questions with the second batch.
>
> Carlos: Will do!

We had similar exchanges for a few times and I wasn't getting anywhere. Fighting this battle long distance proved to be a challenge. My wireframes, after all, were extremely bare boned and this design made the website look like an internal effort instead of a public-facing new product. This, however, was of no concern to anyone else involved mainly because of the tight deadline and the fact that the functionality and features were so much more important. I decided to try once more with Carlos.

> Marina: Hey Carlos, good morning! So I was hoping we can go over the landing page screenshot one more time.
>
> Carlos:
>
> Hello, hello! The landing page screenshot? We're already on the third batch because Yury said he needed it by tomorrow. I sent Laura an update this morning.
>
> Marina: Right, the design is looking good but I'm wondering if we can spruce it up a bit.
>
> Carlos: Spruce it up?
>
> Marina: Yep, like add more design elements? Perhaps a graphic or two?
>
> Carlos: Oh, I see. Let me pull up the wireframes and see what you mean.
>
> Marina: It's not *in* the wireframes because that's not what they're meant for. Can you recommend some layout enhancements?
>
> Carlos: Layout enhancements? I'm not sure what you mean. Our designs look exactly like what you had wanted. That is what you wanted, isn't it?
>
> Marina: Yes, but ...

I didn't know what else to say. I *wanted* to tell them this wasn't acceptable for a final design and was a rough draft at best, but I wasn't finding the right words without being overly direct and, well, rude. It wasn't my job to be dictating the design and I was still recovering from those uncomfortable conference calls that at times felt like a downright barrage on my work. I didn't want to place Carlos in that same situation and so I backed off.

> Marina: Thanks, Carlos. That's fine.
> Carlos: Great, talk to you soon!

And so we limped toward the completion of the project getting increasingly grumpy with each other. Each presentation now started with me habitually on the defensive. I was always expecting push back on my concepts and would begin each presentation in an unhappy monotone. The social commerce team, in the meanwhile, completely mobilized what I had initially dismissed as futile efforts. They held multiple brainstorming sessions and developed a great rapport among a cross-departmental team. Everyone appeared revitalized by this new initiative and some creative concepts started to surface. Pebble, the internal ersatz Twitter, was now peppered with the team's inside jokes. Sure, a lot of their concepts were raw and a bit too high reaching for an immediate practical application, but the entire team was working toward a common goal and each member was aware of that goal.

My project, on the other hand, was at once fading into oblivion and escalating into crisis mode. The oblivion came from the fact that most of the management was so involved with the social commerce project that my project was no longer on their radar. However, the deadline was in sight and I was rushing through the rest of the wireframes after many compromises with Sandip and the California team, usually simplifying the features and functionality in favor of a quicker launch.

CORRESPONDENCE WITH BELARUS

Just when I thought things could not go any worse, the real fun started when another international location was thrown into the mix. We were about four weeks from the release date and I figured the QA (quality assurance) process on the already-completed portions of the site would start any day. The developers, however, had not yet communicated that the site was ready for testing, so I was caught off guard one morning when I received an e-mail from someone named Olga in Belarus. This was my only notification that the QA process had started, and Olga, being in a different time zone, had a chance to begin testing the site hours before I even knew it was released. She immediately found gaping holes in the interaction flow and proceeded to document them in e-mails. Unlike the typical one-sentence e-mails I usually received from Vlad, Yury, or even Carlos, Olga's e-mail read like an essay and wasn't easily scannable.

> To: Marina
> From: Olga
> Cc: Yury; Vlad
> Subject: Landing Page and Account Setup QA

Dear Marina,

I am currently testing the Landing Page and Account Setup sections. The landing page has one big image and the navigation in the header and footer. The logo is in the top left side. Once a user creates an account and logs in, there is a logout link in the header on the right. If the user clicks the logout link, they are taken back to the login page.

In the use case where a user has several types of accounts, what should happen? The user currently would have to log out and then log back in. They would need to maintain two separate login credentials. I was looking in the wireframes and saw no documentation for a different way to do this without this extra step. I do not think it is advisable for one user to have to go through so many clicks just to be able to accomplish something fairly simple.

Therefore, with permission, I would like to propose a different solution. I think it would be better if we put two links in the header once the user is logged in. One link would take him to his first account and one link would take him to his second account. By switching accounts this way, the user can avoid being logged out and then logging in.

I think this way is better and simpler and hope you will agree. I recommend this change is implemented as soon as possible and I will resume testing the rest of the screens.

Sincerely,
Olga

Olga had a point—although it was not that simple. Like her, I had initially proposed that we allow sellers to manage all their accounts with just one set of login credentials and the account view can be switched once a seller is logged in. Forcing sellers to log in and then log out only to see slightly different pages was problematic, but my recommendation was trumped in favor of a speedy delivery. When I saw Olga proposing a similar solution, my initial reaction was a bit smug. "Ha," I thought, "nice suggestion but it's not going to fly." I was about to respond when I noticed that Yury beat me to it.

To: Olga
From: Yury
Cc: Marina; Vlad
Subject: Re: Landing Page and Account Setup QA

Sounds good. Already making the change.

Thanks,
Yury

What? Now I was definitely angry and even though I had suggested it first, my first instinct was to write back to both of them and explain why it was not a good idea. Our site actually had three types of seller accounts, and each type saw a different set of pages relevant to their account when they are logged in. Some sellers were only going to advertise on our site and then link to their own sites for sales. Other sellers were going to actually sell their products on our site but then ship to the customer themselves. The third seller type, and one which would be introduced in a later iteration, would allow us to sell and ship their products for them. If a seller had two accounts with us, they wouldn't be able to simply switch between their advertising and their selling accounts while logged in. They would need to possess two sets of login credentials to manage their accounts.

I agreed there was a problem to be solved but Olga's solution (or my old solution) wasn't optimal for good user experience either. For example, what will happen once we add the third account in the next iteration and then possibly more later? Will there always be so many account links in the header? And what happened to the meeting where we decided we were not going to do it that way, where incidentally Olga wasn't present? And besides, why was the QA engineer making user experience calls? I had to take a step back and breathe because I was starting to act outside my normal professional self. I took a short break and regrouped my thoughts, before typing up a reply.

To: Olga
From: Marina
Cc: Yury; Vlad
Subject: Re: Landing Page and Account Setup QA

Hi Olga,
Thank you for your input. Your recommendation is certainly valid; however, it does not take into account that within the next phase of this project we will have more seller types. With a seller having at least three types of accounts, the navigation would become too crowded with all the links in the header. At that point, I think we should consider a different approach. One way would be to take the seller to a page that lists all the seller's accounts and they can navigate between them without logging out. This would be similar to having a checking and savings account on a banking site.
In the meanwhile, since the team, along with our product manager, had already decided to launch the logout approach for now, we should go ahead and keep what's already there.

Thanks,
Marina

I took a step back and reread what I had just written. It was longer than my typical e-mails and I realized her lengthy essay-style e-mail was reflected in my response. I felt it got the point across in a professional manner, though, and hit Send.

Olga did not respond given the time difference, but Yury's e-mail was short and to the point.

Marina,

Sandip made this call a few days ago and I already made the change. Were you not aware? By the way, send us the updated wireframes to reflect this.

Yury.

I *wasn't*, in fact, aware because no one bothered to communicate this change to me at the time. I also felt completely undermined because my opinion and input were mattering less and less. At this point, I felt I had two options. I could continue this e-mail communication and fight a cause I no longer really believed in or I could drop this incident and talk to Laura about establishing better levels of communication among the team members. I did the latter and also had a short phone chat with Sandip asking him to keep me apprised of any essential interface decisions. He apologized for the misunderstanding and blamed his busy schedule for letting this slip by. Feeling a bit better, I still got the sense this project was plowing full force ahead and leaving me behind.

A DIFFERENT TAKE

But, then, several days after the e-mail exchange with Olga, something curious happened.

As I was leaving yet another frustrating conference call, I ran into Dimitri, the director of development from the California office. He was visiting our offices in Chicago and asked how the project was going.

"Well," I stalled as I considered which answer he wanted to hear.

"You know," he said, "my team tells me this is one of their most promising projects. They're not used to working with a user experience architect and said that it really helps with the design aspect of the site. Sounds like it's going really well! "

"Wow, really?" I couldn't hide my surprise.

"Keep up the good work!" he called over his shoulder as he hurried off to a meeting.

I was shocked. After all, each conference call we held was an exhausting battle. I never had any positive feedback, everything I proposed was hotly debated, and decisions were made without my input. I thought about how nice it was of Dimitri to provide positive reinforcement and wished his team was a little more like him. And then it dawned on me; he was just conveying what his team had told him and they thought this project was going well. Could it be that I had discounted the unique cultural context of our project? Could it be that I had been so accustomed to diplomacy and business-speak in face-to-face talks that I was lost in the shuffle of phone conferences, web meetings, e-mails, and Skype sessions on this project—so lost that I had forgotten, even though it was right under my nose, that this project had a multicultural, globally dispersed team who might have different time zones, cultural expectations, and business norms?

You may wonder why it took me so long to arrive here especially since, and I probably should have mentioned this sooner, I, too, am Russian. Perhaps this is why

I took each interaction so personally. I was born Russian but grew up in the United States and am very used to the American polite way of mitigating each communication. This habit is so threaded into the American culture that, although most people are not aware of Penelope Brown and Stephen Levinson's politeness theory (see Recommended Readings), every speaker in the professional world employs at least some of their strategies. According to this politeness theory, speakers can use various politeness strategies to mitigate face-threatening acts. For instance, instead of commanding someone to do something, one may first express thanks, provide an excuse, or apologize in advance for potentially imposing on a speaker or hurting the speaker's feelings.

This theory also takes into account relationships between the speaker and listener in terms of social and power distances. For example, if the listener is a stranger, one might ask for directions pessimistically: "You wouldn't happen to know where the nearest post office is?" If the speaker is a professor, she may use a bald-on-record strategy with her students ("Listen up, class!") without bothering to mitigate risk to her listeners. In a peer-to-peer communication, though, it's common to distance oneself or use a passive voice, such as "It would be great if this problem was resolved today," to avoid sounding too authoritative.

Of all the politeness strategies, the Russian developers unfortunately employed none, except, perhaps, for the bald-on-record strategy. This strategy is typically saved for emergencies (Watch out!), commanding tasks (Do this!), or the aforementioned power distance situation between the speaker and listener. If peers start talking this way to each other, perhaps the only outcome they will achieve is having something thrown at them.

But with my Russian background, I should know that the Russian language has a very direct rhetoric which sounds odd when placed in an English language context. Indeed, it maps closest to that dreaded bald-on-record politeness (or would it be impoliteness?) strategy. For example, while it is common in the English language to answer the phone using the polite offer "May I help you?" it is not outside the social norm in the Russian language to simply state "And what do you want?"

Olga, the QA administrator from Belarus, whose native language was also Russian, may have been aware of this disparity in politeness between the two languages, so she took great care writing long, detailed e-mails and to propose solutions with the "I think it would be better to..." statement. Or it may be that she was communicating from a different time zone and was in no hurry to write those e-mails. Or maybe it was the fact that she was a woman?

The Russian developers from California, apparently, took the opposite approach. Maybe it was because of their lack of awareness of the language difference, their personalities, their work role, their gender, or simply the stress of the looming deadline. Whatever the reason, there was no small-talk and no "Wouldn't it be better to..." statements, only the "We do this" responses.

As long as I'm talking about culture, I should also mention that Sandip's nationality is Indian. But I felt like his cultural background never figured into our interactions. While his communication style was generally very direct, he was also tactful. I wonder if this might be attributed to him being acclimated to the American culture or him being more of a business than technology person? Sandip and I also saw each other face to face, so it might be that I could see that he was friendlier in those moments than when he was on the phone. I wonder, if I were able to see the Russian developers in person, whether my impression of them would change.

As for the Colombian designers, I cannot say for sure that our interaction was as much culturally charged as it was organizationally so. They were, after all, an external agency hired to help our team during crunch time. Their role was not properly defined as graphic designers, so they assumed the part of production workers by simply taking my wireframes and converting them to Photoshop composite files. After hours of being on conference calls and listening to the rest of the team members verbally spar, they, as hired help, probably had no choice but simply agree to create whatever it was we wanted. And fast. Since the wireframes were so battled over already, they probably decided that they wouldn't want to change a thing about them and would instead insist on assuming that those wireframes were what we wanted. Of course, not being in the same location made it all the more challenging.

AND SO ENDS THE FIRST ITERATION

The first iteration's deadline was met and the website was launched. We worked collectively, more or less, to achieve this goal and walked away with lessons learned. Perhaps harboring resentment was not one of the lessons because apparently the Russian developers did not figure it went as poorly as I did. Indeed, as the first iteration deadline was nearing the end, some of the developers sent us Facebook and LinkedIn invitations. Apparently, they had, in the end, decided we should socially network outside the project.

And I also noticed a slight shift in our conference calls. At one point, I used a new software application, Balsamiq, for prototyping instead of the usual Microsoft Visio. I was working on a new section of the website and wanted to run some concept ideas past the team. Balsamiq rendered the page in a sketchy cartoonlike interface that made the design appear friendlier. It also made my prototypes look like the initial sketches they were meant to be.

"This looks good, Marina," commented Vlad.

"What program were you using to create these? It's neat," the otherwise gruff Yury actually showed enthusiasm.

Was it my imagination or was this interaction friendlier, warmer somehow? To be sure, the rest of the meetings still consisted of them picking apart my concepts with the occasional sigh, but I felt lighter, more receptive to criticism, and more patient with explaining my ideas and standing by my point of view. At least now, we were used to the process and learned the pain points along the way. Knowing how to read a wireframe and how to distinguish between an initial draft and a final version helped speed the rest of the project along with less bumps.

For the rest of my time on this project, I asked Laura to schedule actual face-to-face time with Sandip prior to each presentation. This gave me a chance to hear the business requirements from his point of view and to share my ideas with him prior to presenting them to the larger team. This also allowed us to touch upon on any changes he might have been discussing with the developers so that I could document them as well. This face time proved so invaluable that months after I left the project after the third iteration, Eli, the user experience architect who replaced me, finally received the budget to fly out to California to meet the developers. He later told me that spending an hour in the same conference room with the developers was worth days of phone conferences—and guess what, the Russian developers indeed did not come across as gruff in person!

Months later, a blog I read referenced a marvelous quote commonly attributed to Johann Wolfgang von Goethe:

> I have come to the frightening conclusion that I am the decisive element. It is my personal approach that creates the climate. It is my daily mood that makes the weather. I possess tremendous power to make life miserable or joyous. I can be a tool of torture or an instrument of inspiration, I can humiliate or humor, hurt or heal. In all situations, it is my response that decides whether a crisis is escalated or de-escalated, and a person is humanized or de-humanized. If we treat people as they are, we make them worse. If we treat people as they ought to be, we help them become what they are capable of becoming.

Or, in other famous words, be the difference you want to see. I could not change the way each team member chose to communicate but I could definitely reframe how I interpreted the communication. If everyone on the team had read and heeded this wise statement, our interaction could have been smoother and more pleasant overall. It is sometimes worth considering which battles you pick and what is being traded for the win.

I was reassigned from the project and, in fact, switched companies shortly after the third iteration. Although each iteration went smoother, I was happy to move on. Eli, too, moved on to different endeavors and was replaced by Gabriela, a new addition to the company. Each new team built on the mistakes and successes of the previous and the website is now live and has brought the company the desired revenue.

While shopping in a grocery store months after I had left the company, I ran into Gabriela, in the assorted nuts and grains aisle no less. Curiously, while the user experience role kept shifting from person to person, the IT team members remained the same. We reminisced about the project and the team and why it was that the more creative roles could not hold on for longer than a few iterations.

"Oh well," she said. "You learn and move on. At least the site is live and functioning."

We said our good-byes and, as I went back to browsing the aisle, I thought back to the team and the project. Sure, the site was live, but in layers deep beneath the architecture and the design lay the blueprints and the code that made it all come together due to the efforts of a large team. Each module on every page was carefully constructed and, just like bricks in a building, embedded the history of the process that put it there. That this process was now easier due to the foundation we all laid was certain and, regardless of how challenging it was at the time, I felt wiser for having taken part.

RECOMMENDED READINGS

1. To learn more about politeness theories and their implications for business and technical communication, refer to the following sources:

   - P. Brown and S. C. Levinson, *Politeness: Some Universals in Language Usage.* Cambridge: University of Cambridge Press, 1987.
   - L. Rodman, "You-attitude: A linguistic perspective," *Business Communication Quarterly,* vol. 64, no. 4, pp. 9–25, December 2001.
   - E. B. Brockman and K. Belanger, "You-attitude and positive emphasis: Testing received wisdom in business communication," *Bulletin of the Association for Business Communication,* vol. 56, no. 2, pp. 1–5, 1993.
   - K. Riley and J. Mackiewicz, "Resolving the directness dilemma in document review sessions with nonnative speakers," *IEEE Transactions on Professional Communication,* vol. 46, no. 1, pp. 1–16, 2003.
   - J. Hagge and C. Kostelnick, "Linguistic politeness in professional prose: A discourse analysis of auditors' suggestion letters, with implications for business communication pedagogy," *Written Communication,* vol. 6, no. 3, pp. 312–339, 1989.
   - A. N. Shelby and N. L. Reinsch, Jr., "Positive emphasis and you-attitude: An empirical study," *Journal of Business Communication,* vol. 32, no. 4, pp. 303–326, 1995.
   - P. S. Rogers and S. M. Lee-Wong, "Reconceptualizing politeness to accommodate dynamic tensions in subordinate-to-superior reporting," *Journal of Business and Technical Communication,* vol. 17, no. 4, pp. 379–412, 2003.

2. Usability has grown into an important field of study in the United States; however, advocating user experience is a challenging task in a techno-centered society, where making technology work is often more visible and more important for business. Advocating for the users becomes more challenging when the user group is diverse. The following sources discuss usability studies across cultures, international differences, and implications for usability experts and technical communicators:

   - E. M. Del Galdo and J. Nielsen, Eds., *International User Interfaces.* New York: Wiley, 1996.
   - T. Fernandes, *Global Interface Design: A Guide to Designing International User Interfaces.* Boston, MA: AP Professional, 1995.
   - P. A. V. Hall and R. Hudson, Eds., *Software without Frontiers: A Multi-Platform, Multi-Cultural, Multi-Nation Approach.* New York: Wiley, 1997.
   - F. Zahedi, W. V. Van Pelt, and J. Song, "A conceptual framework for international web design," *IEEE Transactions on Professional Communication,* vol. 44, no. 2, pp. 83–103, 2001.

- F. Sapienza, "A shared meanings approach to intercultural usability: Results of a user study between international and American university students," *IEEE Transactions on Professional Communication,* vol. 51, no. 2, pp. 215–227, 2008.
- S. Dray and D. Siegel, "Remote possibilities? International usability testing at a distance," *Interactions,* vol. 11, no. 2, pp. 10–17, 2004.
- J. Jeyaraj, "Technical communication and cross cultural miscommunication: usability and the outsourcing of writing," paper presented at the IEEE International Professional Communication Conference, Limerick, Ireland, 2005.
- H. Sun, "The triumph of users: Achieving cultural usability goals with user localization," *Technical Communication Quarterly,* vol. 15, no. 4, pp. 457–481, 2006.

DISCUSSION QUESTIONS

1. What do you think were the major obstacles in the seller portal project? What do you think was causing these obstacles: Was it culture, language, time (differences as well as constraints), distance, work roles, gender, all of these, or something else still? Why do you think so?

2. Based on your answer to question 1, how might Marina and others in the story have acted differently to foresee or solve those obstacles?

3. We hear about Marina's frustration in this story. What do you think the Russian developers might have been feeling/thinking along the way? And what about the Colombian designers and Belarusian QA analyst? What could have been their concerns or thoughts about Marina? And how might these different concerns and thoughts be compromised in intercultural projects?

4. Do you think Penelope Brown and Stephen Levinson's politeness theory figures into this story, and if so, how? Do you think this theory can be useful at explaining misunderstandings and conflicts that occur in other communication contexts, intercultural or otherwise?

5. As web and other forms of communication become global, is user experience a more important consideration in global business or is it less so? Can we build something that is truly user friendly when the "user" is globally dispersed and culturally diverse? Can you think of global business communication examples that are and are not user friendly?

CHAPTER 3

LEARNING CURVE

Vaishnavi Thoguluva Vijayaram

Vaishnavi (Vaish) Thoguluva Vijayaram is an independent consultant specializing in technical communication. She has more than seven years of documentation work experience in which she was the sole author for four years. She has been authoring and managing technical documentation projects that include reference manuals, installation guides, user guides (for multiprofiled audience), online help, and training materials on wireless mobile applications, business, banking, GPS-based tracking systems, and accounting products for a wide range of internal and external audiences. Vijayaram is a usability enthusiast who constantly seeks to improve user experience by collaborating with technical, support, and technocommercial teams. She has published on Knol and is an avid blogger who writes on various topics in her blog, www.rattlingcommunicator.blogspot.com. Vaishnavi Thoguluva Vijayaram can be reached at vaishvijay@gmail.com.

CHAPTER SYNOPSIS

In this story, which is told in the third person and based on the writer's personal experience, the spirited, confident Indian technical writer Sajala joined Bear Bhd, an R&D company in Malaysia. Sajala quickly realized that everything she was used to back home in India was now different, from eating habits to communication methods and work environments and dynamics. Sajala made friends, but in her eagerness to make things happen and to push for changes, she also made enemies. The common misconception about the role and work of technical writers seemed to prevail in this Malaysian company. There was also the challenge of proving herself as a capable "foreigner." When things were going well, Sajala was trusted as a writer who has better English than the local Malaysians, but when times were trying, she was considered a "nonnative" English writer who was less competent than U.S. "native" speakers. It was a steep learning curve as Sajala tried to adjust and adapt, all the while trying to do her job.

"Sajala, common yah, you speak so fast, how to follow you? You speak like this at home ah? Or is it just at work, that too only in English?" Mrs. Jega, a friend at OT Software Systems in India, had asked Sajala while she was working there as a technical writer. (All company and character names are pseudonyms.)

With a lighthearted banter, Sajala replied, "Maybe you are not listening when I talk . . . I speak like this at home in Sourashtra* and with some of my friends in Tamil† too!"

How could she have thought that someday the way she spoke would cause her bitter experiences? Sajala is a self-starter, a passionate technical writer with a "can do" attitude. She had never faced communication issues at work while working in her home country, India.

JOINING BEAR Bhd

After working for over three years in India, Sajala, with great zeal, followed her husband, Ishwar, to Malaysia. During her initial days, Ishwar's friend and colleague, Mr. Chandra, an Indian expatriate working in Malaysia for the past seven years, took them to Dr. Baig's Hari Raya Pausa open house. ‡ There, Chandra introduced Sajala to the host Mrs. Shameem Baig, who became her friend and mentor. When Sajala talked about her career, Shameem said, "Maybe you could talk to my son, Shezan. He is a software engineer at SK Management Services."

Sajala was in for luck. After talking to Shezan, she learned that SK Management was in need of a technical writer, although it might not be a full-time job. Sajala did not mind and took up the assignment, her first one in Malaysia. After a short

---

*Sourashtra is an Indo-European or Indo-Aryan language related to Gujarati and spoken by the minority in southern India, mostly in the Indian States of Karnataka, Andhra Pradesh, and Tamil Nadu.

†Tamil is a Dravidian language spoken predominantly by Tamil people of India, Sri Lanka, Malaysia, and Singapore.

‡A significant celebration for Muslims, Hari Raya Pausa marks the end of the fasting season of Ramadan. During the festive season, people have open houses and invite friends and acquaintances to their house for food and refreshment. It is a reflection of the Malaysian multicultural society, where people celebrate the festival irrespective of their religion and race.

stint, she decided to look for full-time positions. She sent her résumé and a self-analysis document to prospective companies, and the first interview she was called for was from Bear Bhd.

Chan, the manager of the Business Process (BP) Division at Bear Bhd, interviewed Sajala. Chan started the interview on a positive note, "Your portfolio and résumé shows that you can help us in organizing our documentation. You are a foreigner, so please tell me how long you intend to work for us, if you are offered with this job?"

Sajala was surprised at this, but she managed to say, "We plan to live in Malaysia until my husband completes his PhD. So, I can work for you at least for three years."

Chan seemed satisfied with this answer and showed Sajala their company's product ecosystem diagram and explained their products. If Sajala was surprised now, she was more so when Chan said, "Now, let me tell about our company and your remuneration. You cannot expect bonus or annual increments because we are an R&D-based company and we run on venture capital funds. We do have U.S.-based clients, so when we complete a project, there will be performance-based incentives and increments. Do you have any questions?"

Sajala could not believe that she was offered a job in such a short time without being asked any technical questions or having any written test. Sajala looked at Chan and asked, "Does it mean that I have a job here?" Chan smiled and said, "Yes, but you are a foreigner, so you will have to go through the paperwork with our HR person, Tanya." Until then, Sajala had thought that job hunting was a time-consuming process which involved several rounds of interviews. Getting a job offer within 30 minutes on an interview was something new for her.

Chan continued to say, "Based on your qualification and experience, I will be considering you as a senior, so I can offer you RMXXXX per month." Here started Sajala's typical habit of comparing anything and everything with its equivalent in India—a habit of hers as a new expatriate. For instance, when she bought tomatoes at the market, she used to convert the price from ringgit to rupee and wonder why tomatoes were so expensive in Malaysia. Well, she was adapting the same theory at work as well. Sajala did a quick mental calculation and converted the amount to Indian rupees. Apparently, Chan mistook that short silence as "hesitation" and quickly raised his offer by 10 percent.

Seeing that there was room to bargain, Sajala commented, "You just said there might not be annual increments. In that case, I am just trying to see if this pay is reasonable for at least two years to come." Chan replied, "Our chief operating officer will be meeting you tomorrow. You can try your luck with him." Rather than luck, Sajala believed in research or "homework." Naturally, she wanted to be prepared for the interview with the chief operating officer (COO). From her colleagues at SK Management and her husband's friend Chandra, she learned that annual increments in Malaysian companies were low and usually ranged from 1 to 4 percent. So, Sajala thought that it would be good if she could convince the COO to increase her pay by another 6 to 8 percent.

The next day, Sajala met the COO Ming in his large and pleasant office. After exchanging pleasantries, Ming started the meeting: "I believe in my managers. I'm sure they make right choices when they select the candidates. Chan says your Ingrish is good." For a minute Sajala was confused, but she quickly regained her composure and smiled, as she remembered that *Ingrish* is the Malay word for English language. Ming continued, "Chan must have told you that we rely on venture capitalists. Are you still interested in working with us?"

Immediately, Sajala said, "Yes I'm interested in working here. The only thing that bothers me is that there are no bonuses or annual increments. I wish you could revise your offer." Ming thought for a while and after doing a quick calculation on his notebook, he increased the offer by another five percent. Sajala was still hesitant, so Ming said, "We will meet up for annual performance review and see what we can do at that time." With that, Sajala accepted the offer.

FIRST IMPRESSIONS

On her first day at work, Sajala noticed that Bear Bhd had a very relaxed atmosphere compared to her previous experience at OT Software Systems in India. At OT, the kitchen was used mainly to heat up their packed lunches. There was no coffee- or tea-making facility. Instead, a company-leased traditional chaai wala (tea vendor) served coffee/tea to the staff twice a day, right at their workstations. To take an additional break, one had to sign in a "movement register." Though it was not seriously monitored, employees took breaks sparingly. However, at Bear Bhd, it was a different story. There was a well-stocked kitchen with tea/coffee-making facility. The staff took breaks as often and when they pleased. There was no monitoring.

Eating out was a part of the Malaysian culture. Though Sajala loved to eat out, she had fewer options because she was a vegetarian. Whenever she asked for vegetarian food at Malaysian restaurants, the waiters would ask, "Can add sea food or egg?" Sajala would say "No." But still, she wondered, "Will they wash the wok to prepare my dish or just use the same wok that was used to prepare non-veg meal?" To avoid all these issues, Sajala would bring a packed lunch from home.

During lunch break, which was between 1:00 and 2:00 PM, Sajala noticed that the staff went out in groups. A large group of young people passed her workstation and Sajala thought that they might be junior business analysts (BAs) and system analysts (SAs).

One of the five females in the group, Pei Chin, approached Sajala, introduced herself as a BA, and asked, "We are going out for lunch lah.* Do you want to follow us or not?" Sajala said, "I'm a vegetarian, so I've brought my lunch, thank you." Pei Chin asked, "Oh! You cannot eat sea food and egg aso†?" Sajala nodded in agreement, so Pei Chin said, "Next time we'll go Chinese vegetarian, so you can come lah." Sajala was surprised and asked, "Are there Chinese restaurants that serve only vegetarian, you mean without any meat?" Pei Chin smiled and said, "Lots of Chinese are Buddhist mah, some don't eat even onion and garlic. Maybe tomorrow we can go for it?" Sajala happily agreed.

---

*"Lah," "leh," and "mah" are used in Malaysia like "yah," "yar," and "nah" are used in India at the end of a sentence in a casual chat.
†"Aso" is a commonly used slang for "also" in Malaysia.

The first few days nothing much happened as the boss, Chan, was on sick leave. She learned that Chan often went on long sick leaves because he had severe arthritis. The acting person in charge of the team was Har, a soft-spoken guy and the most senior BA. Har e-mailed Sajala to get in touch with Lipi, a female Malaysian Indian technical writer on the BP team and to learn about documentation at Bear Bhd. Sajala expected that Lipi would become her first friend at Bear Bhd because they shared the same cultural background.

Next day, when Sajala saw Lipi at her workstation with her coffee mug, meaning she was having her unofficial "tea time," she walked up and tried to casually chat. "Hi, Lipi. Har asked me to discuss Bear Bhd documentation with you. Can you please help me get started?" Lipi coldly replied, "I'm busy with project tasks and I have a deadline. You just go through the standard company process and procedures documents." Sajala tried to ask, "Can I help you in the project?" Lipi bluntly said, "No, you are new here, so you can't work on a project yet!"

Sajala was taken back by Lipi's cold and distant reaction, and again, she thought about her initial days as a technical writer in India, where she was thrown in a project after only a week's orientation. At OT Systems, when asked about something, people never said, "I don't know." Instead, they always made an effort to find out; nobody flaunted ignorance. Whereas at Bear Bhd, the staff always said "Dunno lah." "What a casual attitude toward work!" wondered Sajala. To keep herself busy and learn about the in-house products, Sajala tried to go through the help files in the in-house applications on her PC but these did not help her. It was becoming tedious and boring, as there was no one to answer her queries. They either were too busy or did not know how to explain things to her.

POP-UP MESSAGES

In her attempt to learn about the company and blend in with her colleagues, Sajala found a unique form of internal communication. Instead of having face-to-face conversations, staff at Bear Bhd used an internal chat applet called pop-up messages. In the first week, she got her first pop-up message from Elsie, a senior BA. The pop-up was sent to all BP members.

Elsie > Hi All, Few of us thought of giving ang pow to Lai's son. Interested to join in? Please give me the money by end of today.

Sajala was curious and sent a pop-up to Pei Chin

Sajala > Hi, Whoz Lai and whatz ang pow?

Pei Chin > Lai is senior SA, hez that tall guy sitting at workstation next to the printer. Ang pow is giving money, they r planning to give ang pow for Lai's new born, u wanna join or not?

During her lunch break later that day, Sajala asked Pei Chin what the pop-up was used for. Pei Chin said, "We use it to vent mah, we all 'talk' using pop-up lah." Sajala grinned, "Um, that means 'gossiping,' right?" and Pei Chin giggled, "Not bad lah you, next time I will put you in our thread, you too can 'see' what's happening!" Ho, a junior BA, joined the conversation: "Yes, yes you will know everything, like who is 'really busy' and who is 'pretending to be busy,' who has the latest gadgets and its price and also a great deal about grooming!"

Having learned about this internal communication applet, Sajala tried to message Pei Chin, Ho, and Elsie about how to use the in-house products:

Sajala > How to use d application? no one helping and many links in help are broken!

Elsie > Hey, always big bosses say "help files not helpful." What to do?

Sajala > Is it? Let me see if I can change that!

Ho > Tough, good luk, hehehe . . .

Pei Chin > Try lah . . .

Sajala > Where to find documentation source? Am sure Har and Lipi too busy to answer me . . .

Elsie > On the network, go to, H:\Archive\Wizard

Sajala > Wizard, magic?!

Elsie > Last time, here they called tech writers as wizards!

Sajala > Y?

Elsie > Dunno lah, maybe ask Lipi :P

When Sajala looked at the documentation source, she was stunned. More than 1000 html and image files were stored in disorganized, multilevel folders. Sajala turned back to chat on the pop-up:

Sajala > So many files! totally disorganized . . . no wonder there are so many broken links . . .

Pei Chin > So, what u want to do?

Sajala > Any idea how this source is maintained?

Elsie > When they start new project, you should take the source from the back up of the last project . . . from there u go . . .

Sajala > No version control or any other validating system???

Ho > I dun think so, heard only for application source codes they have some "check-in/check-out" procedure . . . not for doc . . . but confirm it with Lipi

Elsie > Hey, 1st u try to send e-mail to Chan, what you found/think abt doc source. When he come back, you see what he says. but my guess, he will ignore . . .

Sajala was not disheartened. Being a proactive person, she compiled her observations and suggestions and e-mailed them to her boss and the BP team. She assumed that this mail would be an initiative and at the same time hoped to show that she has been doing something useful. To her dismay, there was no response to her e-mail from her boss or any of the team members. Sajala waited for two days, and when there was still no response, she sent a pop-up to her friends:

Sajala > Why no one reply my mail?

Elsie > You sent to Chan and only CC to team mah!.

Sajala > So?

Ho > Here we normally dun reply to mails if it is not directly sent to us . . . why get involved unnecessarily!

Sajala learned an important lesson!

MEETINGS, MEETINGS, MEETINGS

It was more than a month and Sajala was still doing little. She told her friends, "I feel bad for getting paid. No one lets me do any real work. All I'm doing is *learn, find, suggest, mail,* that too without any response!" Pei Chin tried to cheer her up, "Enjoy your 'honeymoon period' leh. Soon the boss will be back. Then you will be overloaded and will have no time to talk aso." Sajala was not happy to learn that people behaved differently when the boss was around. "Not a good sign," she thought. "Why doesn't he delegate his duties when he is not around!" In India, although Sajala was the only writer, people worked in teams and shared responsibility, so the presence of a boss was not the determining factor for work allocation.

By now, Sajala had found that the company had management meetings on Wednesdays. Additionally, on the first Wednesday of every month, the first half-hour of the management meeting was a staff meeting and everyone attended. Before Sajala attended her first staff meeting, Sajala's friends prepared her with what to expect at lunch: "All big bosses talk nonstop, about 'lessons learned,' and we all pretend to listen," said Elsie. Sajala asked, "What's 'lessons learned'?" Ho said, "Oh! The mistakes done on a project become the 'lesson learned' and we are supposed to learn from the mistakes." Sajala replied, "That's good. At least they try to take action." Pei Chin laughed and said, "Wait and see when you attend the management meeting today."

Management/staff meetings were held in the conference room. At her first meeting, Sajala was disappointed with the seating—or rather standing— arrangement. The COO, team managers, and some senior managers were seated around a conference table and faced a white board. The entire staff, around 50, were standing behind them. In fact, the majority, including Sajala and her friends, were standing right behind the COO! The management team discussed a bug found in the latest released version of the in-house product. It was a result of a technical miss-out* during the analysis-and-design (A&D) process. Ming was listing items that SAs should look for while designing a new feature. The discussion was getting technical and Sajala was not able to follow, but she did see that the problem was caused by Lai, who was keeping cool and not trying to defend himself or even apologize.

*Miss-out is a internal jargon for "mistake" at Bear Bhd.

Sajala was still thinking about that miss-out after the staff meeting. She was rather intrigued with Lai's attitude, so she sent a pop-up to her regular group:

Sajala > How come, Lai never reacted. It was his mistake, right?

Elsie > Yes, but heard that he told his boss oredi and was adviced to take a short-cut bcos do full impact study wud be time consuming, so boom this defect came!

Sajala > I C!, but his boss didn't support him, why?

Elsie > Sure no lah . . . no boss will accept his mistake.

Sajala > then what is the point in this discussion that too in a management meeting? What a waste of time!

Ho > Here like that only, can waste "wholesale" company time . . .

Pei Chin > Slowly, u'll learn . . . dun think too much ☺

Elsie > No need to listen what they say, jus be there mah . . .

On Thursdays there were division meetings. Division meetings were held in a discussion room, which had two large desks and each desk could seat six persons comfortably. There were 10 members on the BP team, so people used stools and everyone managed to fit around one desk. Har, the temporary division head, updated the team with news from management meetings followed by status updates from the team members. The division meetings ended quickly and everyone received the minutes later.

Chan came back when Sajala was in her second month at Bear Bhd. On his return, a division meeting was held. At this meeting, Sajala noticed some changes in the seating arrangement. Some of the members preferred to sit behind Chan instead of squeezing around the table. From the beginning, Chan was busy jotting down something on his pad. Sajala assumed that Chan must have been taking notes just for reference. As Sajala was sitting directly opposite to him, she had an opportunity to see what he was "writing" and was shocked to see that he was doodling! Sajala tried to get Pei Chin's attention, but she pretended to be staring intently in some other direction. Then she noticed that everyone looked serious and no one made eye contact. It looked as if they were all trying to safeguard themselves against something terrible!

Following a standard reporting protocol, senior team members reported their work first and others followed. Throughout the meeting, Chan was constantly smirking and he had negative remarks for almost everyone. When someone had very little to report, he commented, "Looks like you have spare time; you should be picking up new skills by now." And when Joshua reported what he had learned besides his project tasks, Chan plainly said, "I don't think this is going to be useful for our current project."

When it was the junior BA Li-ying's turn, she started to explain the issues she faced while preparing a test-case document which was meant to be used by the quality assurance (QA) team. Chan cut her short and said, "What is there to report when that matter is settled?" Li-ying tried to open her mouth again, but Chan gave a stern look and said, "No need to talk so much here. Next." It was like slapping Li-ying on the face! Sajala saw the hurt in Li-ying's eyes, but she did not know how to support her. She scanned the whole room, and everyone seemed to be oblivious of the matter or didn't want to get involved.

When Sajala reported her findings and suggestions to improve documentation, Chan commented, "I don't think we can do all that." Sajala replied, "But, it is a one-time task that can significantly reduce the error rate." Chan seemed eager to brush her off and said, "You are new and don't understand how things work here." Sajala wouldn't budge, "The messy folder structure is the main reason for the broken link errors." Chan simply said, "We will fix them on a need basis, no need to spend time for this kind of maintenance task." When Sajala tried to open her mouth again, Chan snapped, "We are underresourced. Just skip it." To Sajala, Chan seemed to be a very different person now. Gone was the soft-spoken person who had interviewed her.

Though she was disappointed, Sajala decided to wait for the right opportunity to broach this subject again. But Chan followed up with her first. Later that day, Sajala received an e-mail from Chan that stated, "Meet me in the discussion room at 5 PM."

Puzzled, Sajala sent a pop-up to her friends:

Sajala > Boss asking me to meet him @ 5 PM?

Elsie > Oh! mentoring session . . . try to keep cool lah . . . you r in for a long advise :p

Pei Chin > Better go to the discussion room first n wait for him, if not he'll say u r late!

Sajala went to the discussion room five minutes earlier and sat facing the whiteboard. Chan entered and closed the door behind him and walked up to sit opposite to her. He noticed the confused expression on Sajala's face and wryly commented, "Here, this kind of closed room meeting is common when we want to discuss confidential matters." Without waiting for Sajala's reply, he continued, "Sajala, you have to first compete with Lipi and prove yourself before you become a senior, and only then can you suggest any major changes like what you said in the division meeting."

Sajala was perplexed, "Didn't you consider me as a senior when you interviewed me? Now, I'm not sure what you mean by proving myself." Chan acted surprised and said, "Is it? Only for fixing your pay, I considered you as a senior! To be treated as a senior, you have to compete and perform better than others, so it's up to you to prove yourself now."

Sajala wondered if Chan had also spoken to Lipi about competing with her, maybe even before she joined the company. That might be the reason for Lipi's cold shoulder. Always preferring to work in harmony with her colleagues, Sajala said, "I'd rather coordinate and collaborate. I thought that our aim is to come up with quality documentation without competing."

It was Chan's turn to be surprised, but he managed to say, "Sajala, you don't seem to understand how things work here. Maybe you can start fixing minor cases in this project. I will talk to Lipi about this." The meeting ended abruptly and Sajala thought she might have rubbed Chan the wrong way by disagreeing with him.

The next day, Lipi had a "closed door" meeting with Chan, after which she started assigning Sajala with cases that involved fixing minor errors like broken links and missing images. Sajala took up the cases as she was aware that Chan was behind those assignments. But after fixing a couple of cases, Sajala tried to talk to Lipi about the disorganized folder structure, which was the root cause of the errors. "I agree, but remember how Chan reacted in the division meeting? I wouldn't push for it," said Lipi.

Sajala sensed that Lipi was more receptive, so she tried to use this opportunity to break the barrier: "Ok, let me ask you some other thing." Lipi was listening with a smile, so Sajala continued. "I heard that tech writers here were called wizards! Do you have any idea why?" At this Lipi started laughing and said, "Even now they expect magic from us. Wait till you get a case where they want you to cover up a stupid system behavior with a super-duper 'message.' Really, you will have to do magic."

DOCUMENT RESTRUCTURE PLAN

Within a short period, Chan was off on sick leave again. During this time, COO Ming sent an e-mail to Lipi and Sajala (CCed to Chan): "Lipi/Sajala, May I know why we get so many complaints from our U.S. customers on documentation defects, mostly broken links and missing images?"

While Lipi preferred to stay out of this, Sajala decided to use this opportunity and e-mailed back: "I have been fixing most of these complaints and it is like *fire fighting*. Can I meet you to discuss this?" Ming replied, "Yes."

On the same day, a meeting was scheduled between Ming and Sajala at 3 PM in Ming's office. Sajala was meeting Ming for the first time after joining Bear Bhd, so she was anxious. Unlike Chan, Ming had a welcoming smile and started the meeting after some initial pleasantries, which made Sajala relax.

"Ming, the broken links and missing images are mainly coming up because of our messy folder and file structure in our documentation source. We need to clean up the structure," said Sajala.

Ming asked, "So, what is stopping you?"

Sajala replied, "I've already discussed with Chan, but he said that we are underresourced."

Ming answered, "I'll check that with him, but do you have any suggestions to fix this issue?"

Sajala explained, "Yes. We need to restructure the folders/subfolders and group the html files based on the functions of our application. For example, all the topics related to the accounts receivable module should be placed in one folder."

Ming considered what Sajala said for a moment and said, "I can see what you are trying to say, but before I comment, I want to see a workable suggestion. You are new here, so take time to study and come up with a proposal, say, in a month's time." Sajala agreed with a smile.

Chan was back to work two days after Sajala's meeting with Ming. When he saw Sajala walking past his office with her morning cup of coffee, he stopped her and grudgingly commented, "Why did you meet Ming? Sajala, you should have discussed this with me in detail before elevating this issue. You are not loyal to me." Sajala kept quiet as she was not sure how to respond. She sent a pop message to her friends:

> Sajala > Boss expects me to be loyal! where does loyalty come in when he was not even willing to listen when I first approached him?
>
> Lipi > Hahaha . . . thatz y I kept out of it . . . always its like this . . . just do your work alone, no need to suggest lah . . .
>
> Elsie > Lipi is right . . .

Determined to ignore what Chan said, Sajala came up with a proposal for restructuring the document source. Since Chan had come back from sick leave, Pei Chin advised Sajala to e-mail the proposal to both Ming and Chan. Sajala replied, "He didn't like it when I discussed this with Ming. If I mail both of them he will be very angry." But Lipi agreed with Pei Chin, "Sure, he won't like it, but never mind, this is for your safety only. You don't know what Chan might say about your progress if Ming asks about it at the management meeting."

More than a week had passed since Sajala had e-mailed her plan to both Chan and Ming and there was no response. She noticed that the bosses were constantly in meetings and was hesitant to follow up. Eventually, she learned that the endless high-level meetings were about a new project. During that week's division meeting, Chan announced that the new project was top priority and all other projects were on hold, including Sajala's restructuring plan.

By now, Chan had stopped communicating directly with Sajala and assigned tasks to Sajala through Lipi or other colleagues. Chan regularly brushed aside Sajala's suggestions without even weighing the pros and cons. Sajala sensed that she was being ignored and was treated like a low-grade clerk, which made her feel as if she was nonexistent.

When Sajala tried to break this cycle, Chan snubbed her: "This is my way of doing things, and you don't seem to understand how to interact with your superiors." The term "superior" was something Sajala had never associated with management. It had a negative tone for her and implied that the subordinates were somehow inferior. Usually, she was able to leave work-related thoughts at her doorstep, but lately she was unable to do that. Everyday, after going home she had difficulty pushing away the unpleasant feelings. Every morning she dreaded going to work and often was tempted to quit.

WRITING HELP TOPICS

This cycle was broken when Lipi went on annual leave and Sajala got an opportunity to write a new help topic. According to the company workflow, when writing help topics, the technical writer should get information from technical specifications (specs) prepared by the BAs and SAs. Predominantly, the BAs and SAs at Bear Bhd were of Chinese origin and from childhood had spoken primarily Chinese. Therefore, it was natural for them to think and communicate in Chinese, which made it challenging for them to write specs in English. Often, the specs contained information that could guide the developer, but it seldom had information that could be used by technical writers.

Sajala complained about this to her BA friend Pei Chin, "How can I write *helpful* help, when you guys don't tell the purpose or use of this feature in the specs? At least tell what we have that our competitor doesn't." Pei Chin shrugged and said, "I dunno lah, we were told to design like this." Sajala was amazed, "How can you just follow without understanding?" Pei Chin replied, "It's easy to go along with the boss mah, there's no need to argue, so '*chin chai lah*,' meaning whatever they say is ok!"

Sajala then tried to gather information from Kei Seng, who was a senior SA. He refused to help, "Only BAs are supposed to guide the writer as per the workflow. So, I do not have time to spend with you."

Everybody agreed that the company's help content was not on par with the competitor's, but no one was there to set it right. It was like a jinxed loop which could not be broken! As Sajala could not find help internally, she decided to make use of the information available on the Internet. Regrettably, there was no Internet connection on the writers' PCs. She discussed this issue with Lipi when she came back from her leave. Lipi explained, "I heard that once in the past the entire network was down because of a virus. That is why they use one stand-alone Internet PC."

"This is ridiculous," Sajala said. "There are ways to prevent this kind of attack! And, how come the team leaders have Internet on their PCs?" Lipi replied sarcastically, "They are 'privileged' people lah!" Sajala wouldn't give up and suggested, "Let's ask Chan for Internet connection on our PCs." Lipi rolled her eyes and commented, "You never give up! But there's no point in bringing this up lah. Your boss will just tell you to use the Internet PC." Sajala replied, "Lipi, that Internet PC should be put in the trash. It takes ages just to open the browser!" Lipi suggested, "Maybe try to talk to him in private, instead of bringing it up at the division meeting."

Sajala broached this subject in a one-to-one meeting with Chan, but he gave a lame excuse: "If I give you net connection, then I would have to give it to the BAs as well." This did not seem to yield any result.

Out of frustration, Sajala talked to the marketing team leader, Alex Tan, who used to join Sajala and her friends for lunch. Tan listened sympathetically and replied, "I know how difficult it is to do research without the Internet. Don't worry, at the next management meeting I will suggest providing Internet connection to the writers. A good help topic is good marketing stuff too!"

After the management meeting, Sajala was authorized to have Internet connection. Later, through Tan, Sajala learned that at the meeting Chan said that Sajala wasn't grateful for what she got. "I have not given Internet even to the BAs," Chan said, "but when I approved it for Sajala, she didn't utter a single word of thanks, just had a blank look." Sajala was stunned, "Tan, the Internet is not for my entertainment, so what is there to be grateful for?" Though she was happy that management was supportive of the writers' needs, she was constantly worried that Chan might be angry because she got the Internet connection through Tan's recommendation.

CREATING TRAINING MATERIAL

At this time, BAs Lee, Pei Chin, and Elsie Wong were asked to prepare the contents for User Guide and Training Material. Sajala and Lipi were allocated as support personnel for this project. During project initiation, the QA came up with a workflow based on Chan's advice. Technical writers were not included in the storyboarding discussions, which was the most crucial part of the process. Technical writers were only supposed to edit the BAs' work. Another distressing fact was that the infrastructure team (who was in charge of managing database/networks and procuring software/hardware) was supposed to do the research on the help authoring tools for this project.

When Sajala and Lipi volunteered to do the research, QA Manager JJ refused, mainly because Chan had not recommended technical writers to do this job. Sajala decided to do the research on her own time at home and waited for the right opportunity to talk about it. In the meantime, the BAs prepared the contents and sent the completed chapters to Sajala and Lipi. Sajala had just started reading the chapter when Lipi sent her a message:

Lipi > My neighbour needs a haircut badly . . . shall i do it for him?

Sajala > Permission granted:)

Lipi > SHY

Sajala > Can't help you :p. btw, started to edit tm?

Lipi > Not yet. y?

Sajala > Looks like we need to rewrite the whole stuff!

Lipi > Um, v should be writing it like any help topic.

Sajala > Just look at this sentence . . . I think soon I'll forget how to write . . .

Customer, vendor, and employee names and addresses should be memorized and fill in automatically after they are entered regardless of whether or not they are learned on the fly or input via the vendor, customer, or employee screen. Currently you have to set them up first in their respective section.

Lipi > Come let's put on our wizard's hat, then we might understand it! ^^

Sajala > Sure, am looking for it under my desk, hahaha.

When the edited chapter eventually went for testing, the testers said that the contents were either misleading or incorrect.

Learning about this, Lee sent some angry pop messages to both Lipi and Sajala:

Lee > Hey, u should only be editing, y u change contents?

Lipi > We only corrected your sentences . . .

Lee > But u all have changed the meaning, u should only correct our English mistakes!

Sajala > Thatz what we did!!

Because of this dispute, "BA review" was added in the workflow after "editing" to avoid the unnecessary arguments between the BAs and the editors. Further, weekly meetings were scheduled to monitor the progress. Often Yong, the project management (PM) team leader, or Ming attended the meeting. Yong was the only female team leader at Bear Bhd and was popular among the staff.

BAs still complained that the editors were changing the meaning of their writing and editors argued that they had difficulty understanding the English written by the BAs. This resulted in several rounds of editing and rewriting, which caused major delay in the project. To catch up, the PM asked the team to put in extra effort, meaning "work very late"—a decision that raised many complaints on the pop-up messages:

Elsie > Know nothing on writing, hate it when they force us 2 stay back n write . . .

Pei Chin > I can talk in ingrish as it's free style, cannot write lah . . . too formal style

Lipi > Maybe tell boss lah, our work also will be easier . . .

Sajala > U wish . . . at least include us in yr discussions, then we will know what exactly u r trying to say

Lee > Sajala/Lipi, I'll send pop msg to Chan n ask if u guys can come to d discussions. I will include u both in that thread . . .

Pei Chin > Include all of us lah . . . better still include Yong aso

Sajala > Why bring her in this?

Lipi > Aiyo, u dun understand lah, better get someone his level in the thread as support for us . . .

Lee > Good idea, will start a new thread 4 tis & include chan n yong

Lee > Chan, Can Lipi & Sajala join for training material discussions? This might help them to edit or even they can write the training stuff, bcos we BAs find difficult to write . . . what do u think?

Chan > I am really tired of having to treat you all like kindergarten children. You are professionals—I've told u before that this is not acceptable because writers don't have subject knowledge so no need to include them in discussions.

Yong > Chan, what lee suggests sound reasonable, why don't u consider writers for writing training material?

Chan > When BAs can write specs, why can't they write this?

Lee > Specs is diff, there is no story n some more thatz not for d end user . . .

Yong > You all seem to have diff opinion. We will discuss in the mgmt meeting 2morow n let u know.

Chan must have been upset, as he quickly sent another pop-up message to Lipi and Sajala:

Chan > please come over

Lipi and Sajala stood by Chan's cubicle and waited for him to turn around. With a haughty look, Chan commented, "Both of you seem to have your own agenda and, Sajala, you always want to fight me." Sajala winced and tried to open her mouth, but Lipi stamped her foot and signaled her to keep quiet. Chan continued without noticing their reactions, "This is no longer acceptable and I am not sure whether I will tolerate this much longer. You guys always want your way."

The next day, everyone was eagerly waiting to know the outcome of the management meeting when Pei Chin sent a pop-up message:

Pei Chin > Hi all, chk mail. Yong scheduled progress meeting at 10 AM 2molo. so early, u smell something?

Sajala > Whatz 2molo?!

Elsie > U dunno mah? Lot of us cant say "r" .. its 2moro leh! early meetings means they wanna settle things fast ..

Lipi > do u think we all can convince Chan that only Sajala n I must write?

Pei Chin > I don't think he can totally ignore us rite? . . . if all of us start making noise.

Elsie > We dunno how he'll react . . . bcoz there will be 5 of us . . . if all of us say the same thing, Chan has to do something.

Pei Chin > True also . . . but he may try to down us by telling that v r inefficient . . . or dont have the capacity to understand the management.

Everyone in the team sat on one side of the discussion room and waited for the leaders to arrive. At 10 AM, Chan came in with a stern look and sat on the other side of the table. Right away, he started doodling. It was his way of avoiding conversation, thought Sajala.

Shortly, Yong entered with a pleasant smile, apologized for being late, and sat next to Chan. She started the progress meeting, "We need your cooperation to move forward, so everyone must open up, and let's be objective." Encouraged by this, Lee said, "It is not easy to do the writing within the allocated time." Pei Chin agreed with him. And Elsie added, "Writing specs is itself difficult, but it is for internal use, so somehow we manage. But writing training material is a different story. We are expected to use a complicated HTML editor to write the chapters and formatting is eating up a lot of time." Pei Chin then asked, "Why can't Sajala and Lipi write this like any other help topic?"

Yong asked Chan, "It looks like everybody on the team feels Sajala and Lipi should be writing the training material, and I need to understand why you don't think so." Chan answered, "Sajala and Lipi don't have domain knowledge." Yong asked, "Then, how can they write help topics?" Chan quickly replied, "Our help topics explain features and functions available in our applications, so writers can handle this. Training material should be business based. BAs have this knowledge, so they are the best persons for this project."

Sajala seized this opportunity and said, "Well, our help files are actually not helpful because they are merely menu-based explanations. It is better to move away from this style. To write useful topics or training material, I think a writer should know how to source contents and present it in simple English." Chan interrupted and asked sarcastically, "You don't have business knowledge, then how do you plan to write?" Sajala hesitated and looked at Yong for support. Yong rescued her, "I think Sajala can gather info from our BAs and write. Chan, give her a chance and see if it works." Left with no choice, Chan agreed.

Sajala became the main writer for the training material project and Lipi would be available when her live project was completed. At Bear Bhd, there was no style guide for writing, so Sajala studied the competitors' and other available best practices. Before adapting any style she wanted to confirm with Chan and Yong:

Sajala > Is it ok to adapt narrative style in writing and use screen shots in training material?

Chan > No screen shots, maintenance is high, forget using images.

Yong > What kind of maintenance? Letz meet in the discussion room and talk abt this quickly.

Chan > Meet tomorrow at 5pm.

Yong > Chan, remember we decided to settle tm issues ASAP . . . pls spare some time now.

Chan > Well, right now am busy . . . see u both in 30 mins in the discussion room.

It was a closed-room discussion and Chan started, "Often we enhance our screens, then you need to come up with a new image. The chances of missing out are high." Sajala politely said, "We would be updating the content to reflect the enhancements, so naturally we would be updating the image too." Yong suggested, "Sajala, no need to use too many images; you should consider using images only if it is necessary. Chan, what do you say?" Chan appeared to be thinking for a while and said, "I still don't see the necessity. I assume that the user will have the application screen open when he reads the training material, so I'm not convinced." Sajala said, "For simple screens it is ok. But for complicated screens, explaining becomes simple when we use a screen shot." Yong said, "Sounds reasonable," and Chan had to agree.

Based on the agreed-upon standard, Sajala authored a chapter and asked for Chan's opinion. After making minor adjustments, Chan accepted it. It was tested and then sent to the U.S. client.

Soon, the client e-mailed back: "So glad that you have come up with training material. We want to see more details in it. For example, we have different types of users, ranging from managers and accountants to data entry clerks. At the end of the day, we want our users to learn their part by using the training material."

Chan took this as a complaint from the client and called Yong and Sajala for a quick meeting in the discussion room, just before lunch break. He blamed Sajala: "Your narration is ineffective because you are not a native English writer." Sajala disagreed, "Chan, our clients never complained about narration or style. Our client is not even aware of who has written the training material. I'm sure I've written in simple English, so why are you saying that I am a nonnative writer? "

Chan said, "Our marketing material is written by an American, and your narration is not comparable with that!" Sajala tried to explain, "That is copywriting! That style is not suitable for training material." Chan would not listen, "You write short sentences like a typical Asian." Sajala became defensive, "That's not Asian style! Writing short sentences improves readability and. . . ."

The argument continued and finally Chan said, "You don't know how to talk to a manager and you are highly opinionated and disobedient. You always fight me." At this, Sajala turned red and broke down in tears. She rushed to the washroom to regain her self-control.

CONFLICT RESOLUTION

That was the first and last time Sajala cried in public. She was in no mood to go out for lunch, so Pei Chin and Lipi stayed with her in the office, and Elsie offered to get lunch for the three of them. Realizing that the matter was getting too emotional and personal, Yong scheduled a conflict resolution meeting between Chan, Sajala, and herself at 5:30 PM on the same day, again in the discussion room. This time, Sajala was the last person to enter the room, as she didn't want to face Chan alone before Yong's arrival.

As Sajala entered, Yong was pulling out a stool opposite Chan. Sajala carried a glass of water with her and sat in a corner stool, so that she could see both of them. There were no pleasantries.

Yong questioned Chan first, "You seem to be harsh with Sajala. Can you please explain?" Chan said, "Sajala always fights or argues with me. . . ." Sajala shook her head and tried to open her mouth, but Yong stopped her, "Sajala, let Chan finish. You can explain later."

Chan continued, "During discussions, she is aggressive and interrupts. She always wants to do something big and comes up with far-fetched ideas. When I try to reason with her, she just gives me a blank stare. She doesn't respect me."

Yong turned to Sajala, "Sajala, why do you fight Chan and give blank stares?" Sajala answered, "I never fight or stare. When I try to suggest something, he always says 'convince me.' So I keep explaining. Then without listening fully, he criticizes and refuses to discuss it further, so at times I just keep quiet and look at him. How am I supposed to convince him, when he is not prepared to listen?" The atmosphere in the room was tense. Yong kept silent and Chan was scribbling furiously on his notepad.

After a few moments, Yong turned to Chan again, "Chan, Sajala complains that you never listen, is that so?" Chan replied, "No. She talks fast and uses a lot of tech writing jargons* when she tells me something. It's hard to follow."

"In that case you should have asked her to clarify, but why did you refuse to listen?" Trying to sound authoritative, Chan replied, "All her suggestions seemed to be resource hungry and with our tight schedule we can't entertain them."

Yong was surprised, "You could have discussed resource allocation with the PM. Remember, Ming always encourages new ideas and insists that we should allocate time for research-related activities." Chan gave a noncommittal sigh and Yong con-

---

*"Jargons" rather than "jargon" was frequently used at Bear Bhd.

tinued, "Chan, I agree that Sajala speaks fast, and in meetings, she tries to explain herself in challenging situations. Maybe you took that as interrupting and aggressive."

Turning to Sajala, Yong continued, "And, you should try to talk slowly and make sure that people understand you. Chan does not have a tech writing background, so try to avoid using jargons when you speak, or at least explain it." Sajala nodded in agreement.

Yong then asked, "Have you ever faced communication issues in India, I mean as a writer?" Sajala said, "No. As the only writer, I reported directly to the project manager. He used to listen to my suggestions and asked for clarification if necessary. Here, Chan never asked, so I assumed that I was clear and didn't realize that he didn't understand me. Moreover, he never communicates with me. When he wants me to do something, he passes the message through Lipi or a BA. I feel as if I've lost my dignity."

Chan stopped scribbling and said, "I delegate tasks through Lee to other BAs, so it's not only with you. There is no other intention, it's pure management style. Here no one pushes for things to happen, except for you. You think I can follow your thoughts, but apparently not. My criticisms were meant to be constructive."

Sensing that Sajala and Chan were still wary of each other, Yong tried to appease them both: "Looks like what you two have is partly misunderstanding and partly different expectations. Here most of us don't push for things when they sense their boss is not keen on them. They are rather submissive. But Sajala, you are different, probably because you were used to a different kind of environment. It is not enough to talk slowly. You have to do a feedback loop and make sure that the other party understands what you say. This is because your common sense and my common sense may be entirely different."

Sajala smiled and nodded as she was starting to see why she had to explain even the basics of her profession to the decision makers. As for Chan, Yong suggested, "Your management style seems to be effective with the locals. In Sajala's case, why don't you try to communicate directly with her? Ming once said that Sajala has ideas to improve documentation standards. Coordinate with the PM and allocate time for her to research in that area."

Getting back to the training material issue, Yong recommended, "Let's e-mail our clients and ask for a detailed feedback as to whether they had any difficulty in understanding the training material or the content itself was insufficient."

Within a week, the client sent the feedback. It turned out that they only wanted the training material storyline to be fleshed out with a wider range of user types. They added that the writing was for the most part simple and clear. Sajala was happy to hear this and ready to make more changes. She told Yong and Chan that she could rewrite some of the less clear parts and add more user types. She also talked about other recommendations, citing competitors' writings and some of the best practices on technical writing. For the first time, Chan agreed to Sajala's suggestions without any "convince me" speech.

It took more than a year for Sajala to breathe a sigh of relief and adapt to her new environment. She felt she had changed a great deal and mellowed. She still had arguments with Chan at times, but trust and confidence were slowly growing between them. When she recalled her earlier days at Bear Bhd, she tried to remember not to jump to conclusions that were colored by her experience. As Yong said, "Your common sense and my common sense may be entirely different."

RECOMMENDED READINGS

1. Malaysia is a multicultural society where multiple ethnicities coexist, multiple religions are practiced, and multiple languages and dialects are spoken. To learn more about Malaysia's multicultural diversity and its implication on the various aspects of life in Malaysia, consult the following sources:

   - Z. Abdullah, "Cultural diversity management in Malaysia: A perspective of communication management," in *Managing Cultural Diversity in Asia: A Research Companion,* J. Syed and M. F. Özbilgin, Eds. Northampton, MA: Edward Elgar Publishing, 2010, pp. 14–38.
   - R. Ibrahim, "Multiculturalism and education in Malaysia," *Culture and Religion: An Interdisciplinary Journal,* vol. 8, no. 2, pp. 155–167, 2007.
   - L. M. Li, "A visual analysis of festive television commercials in Malaysia," *Multilingua: Journal of Cross-Cultural and Interlanguage Communication,* vol. 30, no. 3–4, pp. 305–317, 2011.
   - C. Selvarajah and D. Meyer, "One nation, three cultures: Exploring dimensions that relate to leadership in Malaysia," *Leadership & Organization Development Journal,* vol. 29, no. 8, pp. 693–712, 2008.

2. As a result of centuries of colonial rule and western influence, many varieties of English are used around the globe today, giving rise to world Englishes. This idea is changing and challenging the traditional concepts of British Standard English (BrSE) and American Standard English (AmSE). Australia, the Philippines, and other countries are upholding their versions of Englishes, and in places such as India and Malaysia, English is used in several varieties. The following sources provide more information on the concept of world Englishes:

   - M. J. K. Bokor, "Connecting with the 'Other' in technical communication: World Englishes and ethos transformation in U.S. native English-speaking students," *Technical Communication Quarterly,* vol. 20, no. 2, pp. 208–237, 2011.
   - M. J. K. Bokor, "Moving international technical communication forward: A world Englishes approach," *Journal of Technical Writing and Communication,* vol. 41, no. 2, pp. 113–138, 2011.
   - B. B. Kachru, Y. Kachru, and C. Nelson, Eds., *The Handbook of World Englishes.* West Sussex, UK: Wiley Blackwell, 2006.
   - M-Z, Lu, "Living-English work," *College English,* vol. 68, no. 6, pp. 605–618, 2006.

- M-Z. Lu, "An Essay on the work of composition: Composing English against the order of fast capitalism," *College Composition and Communication,* vol. 56, no. 1, pp. 16–50, 2004.

3. To facilitate international and cross-cultural exchanges of information, some scholars argue for the global English approach. This approach advocates using simple, clear, and culture-free forms of English to facilitate the translation process and to help nonnative English speakers understand English-based communication. The following sources provide more information on this approach:

- J. R. Kohl, *The Global English Style Guide: Writing Clear, Translatable Documentation for a Global Market.* Cary, NC: SAS Publishing, 2008.
- E. H. Weiss, *The Elements of International English Style: A Guide to Writing Correspondence, Reports, Technical Documents, and Internet Pages for a Global Audience.* New York: Sharpe, 2005.

DISCUSSION QUESTIONS

1. In what ways does the concept of global English differ from the concept of world Englishes? Which concept do you prefer and why do you prefer it? Who might prefer the opposite of your choice and why might they have that preference?

2. What do you think of the way Sajala tried to "break into" Bear Bhd? What might you have done to prepare yourself for such a transition into a new environment and workplace?

3. What opportunities and challenges does a multi-cultural context such as Malaysia bring to professionals like Sajala?

4. What do you think of the internal communication method, the pop-up messages, used at Bear Bhd? What kinds of organizational culture and workplace dynamics do they help to foster?

5. What do you make of the apparent hierarchical tension at Bear Bhd between the management and the employees, the superiors and the subordinates, and the seniors and the juniors? How was power gained, executed, and dealt with? How does it reflect the larger social and cultural context in Malaysia? What challenges does this hierarchical tension create for technical communicators?

6. Do you think Yong's "conflict resolution meeting" was going to be helpful at solving all of Sajala's problems? What really was causing the conflicts between Chan and Sajala and how might they be solved, if at all?

7. Given the concept of world Englishes, consider the use of English in today's globalized economy and information network. Should there be a standard English or should we celebrate varieties? If a version of standard English is desirable, what standards and whose standards should be adopted? And what do all these mean for the work of technical communicators in the global era?

CHAPTER 4

LOST IN TRANSLATION

Elena Groznaya

Elena Groznaya is a visiting assistant professor at the Graduate School of Business Administration, Kobe University, Japan. Her major areas of research and teaching are intercultural and diversity management. More specifically, Groznaya concentrates on the topics of present-day business cultures of Asia and Eastern Europe with the focus on human resources management of multinational corporations as well as post–mergers-and-acquisitions and employee motivation issues. Besides the academic work, Groznaya applies herself as a freelance corporate trainer, consultant, and writer. Her publications on Indian, Japanese, and Russian business cultures as well as other intercultural business communication topics have appeared in a number of academic journals and management magazines.

CHAPTER SYNOPSIS

This story is based on the experience of the author's Japanese colleague. Three different presentations, two unwittingly confusing presenters, one volcano eruption, and there you have it, a recipe for chaos. Facing this chaos is the hardworking and intelligent Japanese young woman, Yukiko, who was a West-educated sports designer working for a Japanese producer of golf equipment and golf wear. Yukiko's international experience and language skills made her the de facto communication specialist when the company moved steadily into the global market, taking with it its traditional Japanese business practice and custom. To create a global image, the company decided to develop a set of global design guidelines for its products and needed to present this vision to its international subsidiaries. Trying to communicate with the company's international subsidiaries while appeasing her Japanese supervisors, Yukiko was on a rollercoaster of being confident, excited, confused, and worried. The company's British subsidiary staff did not fare any better in their attempt to communicate. It was a hard lesson for all to swallow: Good intentions do not always get results, and carefully crafted messages, unfortunately, easily get lost in translation.

————————

After the official dinner, Simon approached Yukiko, who was sitting at the long table with her colleagues and seemed to be deeply absorbed in her thoughts. Normally, both Simon and Yukiko enjoyed this modern but very cosy Kobe beef restaurant with its warm lights, paper walls, low tables, great food and famous Japanese service. This time, however, neither of them seemed to enjoy it. "Listen, Yukiko. We cannot afford to foul this up. What if something goes wrong!? . . ." Yukiko sighed, looked at Simon and suggested they move to a quieter place in the corner of the room. . . .

JOINING GREEN HILLS

The story started in 2009, when Yukiko was hired by Green Hills, a Japanese company specialising in the design and production of professional golf equipment and golf wear. (All company and character names are pseudonyms.)

Green Hills was founded in Osaka, Japan, in 1983, starting as a small traditional family business. Throughout the years, it maintained a Japanese business structure and was in many ways a typical Japanese corporation. In 2008, however, the company's management came to the conclusion that the local market had offered no opportunities for further growth and decided to go international. The company's management recognised that in order to capitalise on foreign markets, direct presence was needed. By 2009, after a few successful acquisitions, Green Hills had established subsidiaries in North America, Australia, and Europe. In the 2009 fiscal year, the global company generated 30 billion yen in net sales. Sixty-eight percent of the company's income came from the sale of professional golf equipment and 32 percent came from golf sportswear and accessories. Around half of these sales were generated by the foreign subsidiaries.

As a result of the global acquisition and market growth, a large number of non-Japanese employees joined Green Hills. Because they came from different cultural and organisational backgrounds, how to create a common corporate culture and how to establish effective communication between the company's international offices and headquarters became a pressing issue. The management at Green Hills decided to make English the company's official working language. It also started to reconsider other communication and work practices in order to turn the traditional Japanese organisation into a more modern and globalized company while maintaining its

Japanese business model at the headquarters. In the meantime, the human resources department was encouraged to recruit more young people who have international experience. And Yukiko was one of them.

Yukiko Omura was 31 years old. She was born and raised in Osaka. At the age of 9, she went to an English school in Japan, planning to study English literature. But pretty soon she realized that her real passion was fashion design—and sports. Inheriting her interest in tennis from her father, Yukiko had been one of the best tennis players at her school. Wanting to combine her love of sports and her talent in design, Yukiko started making attempts at drawing models in tennis and golf wear.

Yukiko's parents decided that design could be a good job for a girl so after Yukiko graduated from high school, they sent her to the United States to study sportswear design. After staying two years in New York studying for a B.A. in sportswear design, Yukiko moved to London where she received her M.A. in performance sportswear design. She enjoyed her time abroad immensely, but still, she wanted to come back to Japan so she could be close to her friends and family.

Coming back home after her five years spent abroad, Yukiko noticed pretty quickly that she had become "different" from her old school friends. Her liberal attitude and somewhat sporty appearance did not always match the style of her friends, who complained that she was too direct, pushy, challenging and ambitious. At 167 cm, she was also quite tall for a Japanese woman and it made her feeling awkward always being the tallest on the bus or at a party. At her age, most of her friends were either married or about to get married. Yukiko made a few honest attempts to date men, introduced to her by her friends, but she never moved further beyond the first introductory dinner. While frustrated, Yukiko did not want to give up and decided to give herself more time to readjust back to the Japanese life style. She made a few friends among the multinational expatriate community in Osaka. Yukiko felt it was a good solution as she could relate to them and could keep on practicing English. On the other hand, her foreign friends who could manage only very basic Japanese were glad to have a "real but not too Japanese person" among them.

On the professional front, Yukiko was very lucky. It took only two months for her to find a good position with Green Hills. She joined the global design group that was responsible for developing general design directions for Green Hills' sportswear department.

MR. HIRANO'S EXPECTATIONS

Yukiko's new boss, Mr. Hirano, president of the global design group for Green Hills, was in his late 40s. He had been with Green Hills for 25 years. Mr. Hirano's kids were already grown up and lived in Tokyo. His wife was interested in traditional Japanese music and spent a lot of time with her friends visiting theatres and museums and arranging small concerts. Mr. Hirano did not share her interests. He was quite happy working late during the week and playing golf on the weekends with his friends, colleagues and customers.

Leading one of the most creative teams in the company, Mr. Hirano was, however, not the most creative person—indeed, Yukiko quickly found that he was quite systematic, logical and somewhat conservative. In spite of the casual dress code at

Green Hills, Mr. Hirano always wore a black suit in the office. When asked why, he would smile and say, "I do not think a uniform is such a bad idea. I like to look and feel professional. The suit helps me concentrate on my work. Jeans and leisure trousers are good for free time, for weekends and for younger boys. Probably, I am simply too old for such freedoms." Mr. Hirano preferred personal conversations to e-mails and spoke more formally than others. He was tolerant of the "creative" look and behaviour of his subordinates, but as the boss, he tried to keep things within limits. He was ready to listen to employees and offer advice with personal issues but expected loyalty from them and their respect for his position and authority. The team members called him "daddy" behind his back and made regular jokes about his outfit and manners. At the same time, they respected him for his experience and loyalty to the company.

Mr. Hirano had high expectations of Yukiko. He found Yukiko confident, communicative and well-qualified for the job. She was a bit westernised but still familiar with Japanese culture. Mr. Hirano hoped that this young lady, with her professional skills and international experience, could become not only a successful designer but a keenly needed mediator between the company's Japanese headquarters and foreign subsidiaries.

During her first month in the organisation, Mr. Hirano had made this expectation clear to Yukiko. He invited Yukiko to his office and said, "Omura-san,* you have surely noticed that we often experience problems with our foreign design teams. We do not expect them to speak Japanese and accept our way of working, but we hope that they will be able to follow our standard procedures and meet basic requirements. Only last week an Australian team was openly unhappy about the new design requirements we sent to them. You lived abroad and I expect you should be able to understand these issues better than I. We need to improve our communication and make them understand our requirements."

Yukiko had noticed the communication problems in the company. In spite of sincere attempts from all sides to communicate effectively and to do their jobs properly, the level of frustration and disappointment was pretty high. It seemed that although English was made the official language of the new internationalised Green Hills, people still spoke different languages and could not reach understanding. But Yukiko was hesitant to take on this responsibility; after all, it was not an official part of her job.

"Hirano-sensei, would you like me to work on the communication issues as well?" Yukiko replied, "I am afraid it could be a little difficult as I have a great variety of other duties and do have at least eight hours overtime per week."

Mr. Hirano looked at her and said in his quiet, confident manner, "Omura-san, I am quite sure you will find a way to manage it. Maybe you need to be more organised. Please, work harder."

Yukiko understood very well that her attempts to argue or offer more excuses would not change the situation. She could only complain to her friends when they

---

*Japanese use various titles when addressing different people. The "-san" suffix is a generic title of respect and can be used for both males and females. The "-sensei" title used later is a polite form to address someone older, of higher status, and a teacher/mentor to the speaker.

went out that Friday evening, "This man wants me to combine two jobs. I am paid only for one. And I can hardly manage all the duties I have at the moment! How am I supposed to squeeze in something else?"

But by the following Monday she had calmed down, considered the situation once again and even felt somewhat confident about her abilities to solve the communication issues.

Her understanding of both Japanese and Western languages and cultures was her main asset. Mr. Hirano's English skills were not very good and he preferred Yukiko to handle all English communications for him. Within a short time Yukiko took control over information exchanges between the global design group and the regional teams. Because of the time difference between the headquarters and the subsidiaries, she often had to stay in the office until late in the night or work from home. She provided translations, explanations, and other needed information to all sides.

At times she felt exhausted, and Mr. Hirano seemed satisfied but still insistent. "Omura-san," he said once to Yukiko, "I noticed we get less complaints from the designers recently. Let's hope it stays this way. However, there were some misunderstandings with the London team last week. They expressed quite unequivocally their dissatisfaction with our guidelines for the next summer collection. Our team believes the guidelines were clear enough. Probably you could still try a little harder."

"I see," Yukiko said, "Hirano-sensei, I would really like to improve my performance. What would you suggest in this situation?"

Mr. Hirano looked at her, smiled and said, "Just work a little harder and show a little more loyalty."

This did not help much. But as Yukiko could see improvement in the communication with the regional teams, she did not feel frustrated and kept on working as hard as she could.

THE NEW GLOBAL IMAGE

In January 2010, the Green Hills global design group launched a campaign to unify the company's product design. Prior to this, the headquarters only decided on products' technical characteristics and materials, and the subsidiaries had freedom to create their own designs for the local market. Because of this, the look of the golf apparel and accessories varied greatly among the subsidiaries in spite of their shared brand name. The new global brand policy would require the subsidiaries to participate in the development of a unified product appearance proposed by headquarters and based on the needs of the global market. By doing so, the global design group hoped to create a strong, globally recognisable brand image. The freedom so dear to the subsidiaries was expected to be reduced. Moreover, the new policy required even more effective cooperation between design teams from the different regions.

After a few months of group discussions, Mr. Hirano felt that he was ready to present the proposal for the new policy to the executive board. He summoned Yukiko to his office, "Omura-san, I feel we are ready to offer our suggestions to headquarters. I believe you should take this responsibility. Your presentation style is good and, as one of the developers, you are very familiar with our main suggestions. Would you please prepare the presentation? If the campaign is approved by the top management in Japan, your next step would be to inform the subsidiaries about the new policy sometime between April and November. In addition to the new policy, you could deliver the main elements of the global Design Directory your team had developed for 2012. I know this task looks challenging, but I believe you like challenges and you are experienced enough to manage it without major problems. We trust you, but you should remember that this is a very big responsibility."

Yukiko felt the same way. The task was challenging but very exciting. She knew it was her chance to show her abilities and to be noticed by the board.

Yukiko spent the next two days working on the presentation. She wanted it to be clear, logical, detailed and persuading. Mr. Hirano himself sent her a few slides that showed the main steps for the new policy's implementation.

Yukiko shared with Mr. Hirano the finished draft. "Hirano-sensei, as you advised, I created an elaborate presentation. At the moment it has fifteen slides. I think it covers the main steps of the implementation of the policy as well as major concerns and probable solutions. I also included the slides you sent to me yesterday. Thank you for your help."

"Let us go through it together," Mr. Hirano said. He thoroughly checked every slide. "I believe we need to emphasise in more detail the main risks and the solutions. Besides, why not add a slide with the names of the contributors? The slide about the advantages of the policy should go at the end. First, provide all the details. Conclusions, if needed at all, go last. The top management is intelligent enough to make conclusions without your explanations."

Yukiko had to work literally day and night. During the day she did her regular job. At night and on the weekends, she worked on the presentation. She had to go through it with Mr. Hirano at least two times a week and continue to edit it.

On Monday, March 1, came the meeting with the board. Yukiko was nervous and could not sleep the whole night before. She lay awake repeating the main ideas of the presentation in her head. In the morning she felt even more scared. Her mother gave her a cup of tea that was supposed to help her calm down, which did not work. And it was too early, Yukiko thought, to have a glass of wine. Yukiko knew that her career with the company would depend very much on her presentation, the thought of which almost made her feel sick.

She arrived at the main office at 8:30 and the meeting started promptly at 9:00.

Mr. Hirano addressed the board first: "Esteemed president, board members and colleagues, as a leader in global design I present to you some proposals made by my group. We kindly ask that questions be held until the end. I give you Omura-san."

Yukiko thanked the meeting participants and Mr. Hirano and began the presentation.

"Thank you Mr Hirano. We would like to share with you a few ideas." Yukiko opened the first slide that showed the company name. "First of all, we would like to thank you for the opportunity to make this talk and for your support of our ideas."

She then moved to the slides with the names of the contributors who were involved in the development of the new policy (Figure 4.1).

Global Design Policy was developed and approved in 2010 thanks to

- Mr. Horiuchi Kazuhiro堀内 一浩– President
- Mr. ToriiKenta 鳥居健太 - CFO
- Mr. Hirano Ken -平野研– Leader of Global Design Group
- Mr. Matsuo Hiroshi -後藤寛 - Leader of Global Marketing Department
- Mr. Goto Daichi 後藤 大地– Member of Global Design Group
- Mr. Horie Riota 堀江健太– Member of Global Design Group
- Ms. Omura Yukiko 大村 由希子

Figure 4.1 Contributors. As a sign of respect and appreciation, a few slides of the presentation include names of the contributors in hierarchical order. This and the following slides included in the story represent the original slides with their word choices, grammatical forms, and designs.

"Next, we would like to tell you about the functions of the global design group. We create the design guidelines and proposals for the sportswear and accessories. In addition, we coordinate communications between headquarters and the regional design groups. Our leader is Mr. Hirano. You can see from the slide that our group has existed for three years and employs eight people" (see Figure 4.2).

Yukiko switched to the next slide: "Here you can see the functions of the regional design groups and their locations. Our main foreign office is in London" (see Figure 4.3).

 Functions and Goals of GDG

GTG controls design strategy of all parts of the "Green Hills".

Leader of the Global Design Group is Mr. Hirano Ken (平野研)

From 2011 GTG will share high quality of information Globally. It will share amount of information Globally. Global Design Group will develop multiple recommendations and data from the Global Design Office will be provided to many employees by centralized management of Design Center.

By doing this, Global Design Group will provide detailed information to the regional offices and explanations of the new design policies. Help of corporal network to improve the communication between the Headquarters and Regional Design Teams.

Figure 4.2 Global Design Group (GDG) functions and goals. Full sentences and detailed descriptions are used to offer maximally detailed background information and reduce possible misinterpretations.

Functions of the regional design teams and their locations

"Green Hills" has offices and design teams in a number of regions, such as Australia (Sidney), Europe (London, Zurich, Barcelona) and USA.

The regional design teams develop the regional design strategies for the products of our company. They apply their visual designs to the products' technical specifications that comes from the headquarters. They can develop the visual elements and colors for the local markets based on the local market predictions and preferences.

The revenue is distributed like this: USA has 27%, Australia has 22%, England has 31%, Switzerland has 12% and Spain has 8%.

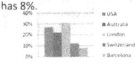

Figure 4.3   Regional design team functions and locations. The slide gives an example of the typical "patchwork" presentation design.

During the next seven minutes, Yukiko offered a detailed description of the company's existing regionally based design approaches. "Once a year in April, the heads of the regional design and marketing teams met in Osaka. The global design group presents the following year's design strategy, including new technical characteristics, materials and features of the sportswear, as well as a new key colour trend. After the global meeting, each regional design team develops its own visual design of the product based on the general suggestions of the global Design Directory proposed by the global design group." She ran through the three slides with the text and schematic images of the existing policy. (See Figure 4.4.)

The next four slides then used schematic images to illustrate, step by step, how the current and the newly proposed policies differ. And Yukiko continued to present: "The new policy suggested by our team asks that the global design group in Osaka develops the combination of key colours and visual design elements that would be compulsory for all regions and all products. This Design Directory would be presented at the global meeting in April. By October, all regional design groups will have to make design proposals for the local markets within the narrow scope of the new global Design Directory. These minor variations should also be approved by the headquarters."

"We would like to offer some possible ways of implementing the new work procedures," Yukiko switched to the next four slides that basically repeated what she had said before.

After that Yukiko moved on to one of the most important topics of the presentation: "Now, we would like to draw your attention to possible concerns and risks connected with the new policy. On this slide (Figure 4.5) you can see some explanations of the expected phenomenon of 'fading intentions.' Our team believes that the

The global design group in Osaka develops the combination of key colours and design elements that must be used in all regions and all products. This Design Directory will be presented at the global meeting in April. By October, all regional design teams will have to make design proposals for the local markets within the offered scope of the new global Design Directory. All variations must be approved by the headquarters by the end of the month. GDG will be offering consultations and additional explanations to the regional design teams concerning the visual design intentions.

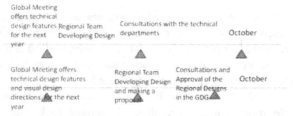

Figure 4.4 Current regionally based design approach. Yukiko does not use key words for her presentation but rather writes full sentences. She hopes such a strategy will help her audience reduce confusion and misinterpretation.

major problems would be caused by this phenomenon when the intentions and the suggestions of the headquarters' team are misinterpreted by the regional offices. The problem could have a big impact and be very unpleasant and costly. This could happen because of poor communication or diverse cultural and educational backgrounds of the designers."

Possible risks policy and suggested solutions for the new design policy 2011

- The new policy of the global visual image design has some risks.

Development of visual design directory by the Global Design Group and sending it to the Regional Groups can cause the effect of fading intention and misintelpretation. This can be caused by cultural differences and educational backgrounds and English skills. And it can be eventually very expensive. The Global Design Group can arrange presentations and detailed explanations for the Regional offices. Effective and clear communication between the GDG and Regional Offices will be very important.

Figure 4.5 Possible risks of new policy. Prevention of possible risks is one of the crucial elements of the new policy. English words with "L" and "R" sounds are often misspelled in Japan, as the Japanese language does not distinguish these two sounds. Hence, "misintelpretation" rather than "misinterpretation."

Yukiko was about to add "as well as natural personal differences" but managed to stop herself. During one of their discussions, Mr. Hirano had been very unhappy with the mentioning of "personal differences" and called it too individualistic and aggressive. Yukiko tried to defend herself and argued that people within the regional groups also have their differences.

"And let them, but when they work together, they belong to one group," stated her boss. "These self-centred attitudes can be really unhealthy and counterproductive."

Yukiko did not insist, but she managed to save one of her favourite examples of the cultural differences: "For example, what a Japanese person traditionally calls blue could be green in other parts of the world. Moreover, English language skills of the designers in the regional offices are not equal. Sometimes they are not good enough to express fine details of the design philosophy."

At this point, Yukiko looked at the board members and Mr. Horiuchi, the quite young and westernised company president. He looked back with curiosity and smiled at her. She felt that everything went well and moved on to the part where she offered some suggestions for handling miscommunications.

The next three slides offered solutions and procedures to prevent the fading of intentions, such as close co-operation and regular consultations between the regional and global design groups.

Finally, Yukiko moved onto the last slides and presented how other large companies offer a unified global design (Figure 4.6). It was not an easy task to persuade Mr. Hirano that these slides were needed, but after a few lengthy discussions, the boss agreed that this information made sense. Yukiko showed the names and images of some of the biggest sportswear producers. "These companies implement the policy of the 'global image,'" she said, pointing to the logos of Nike, Puma, and Lacoste. "This is why their products are well-recognized in any part of the world. We

Conclusion

- Many companies implement the unified design approach. This approach proved to be very effective to produce the global image of the company brand and products. **NIKE   PUMA   LACOSTE**

logo   logo   logo

We will start implementing the new policy of global design direction from January 2011.
Please, find the detailed description of the DD in the offered by the GDG brochure.
Please, follow the new regulations from January 2011.
Please, follow the strict non-disclosure regulations when handling with the brochure.
The Global Design Group members will help you with the development of the local designs based on the Global Design Directory.

Figure 4.6 Conclusion. This slide covers the major concerns: anxiety and uncertainty of the teams involved in the new policy as well as its legal framework. (The NIKE, PUMA® and LACOSTE™ logos used in the original presentation were removed here due to copyright regulations.)

believe the unified visual design could help us make our global image stronger."

With that, Yukiko ended the presentation. She knew that later that day Mr. Hirano would respond to all questions and provide additional explanations to the board and marketing department.

On Thursday evening, Mr. Hirano announced to the team that the global design suggestions were accepted: "I would like you, Yukiko, to continue working on this project. The president expressed satisfaction with your presentation. Your next step would be to deliver the presentation to the subsidiaries. Mr. Matsuo, head of global marketing, has prepared a few additional recommendations for the presentation. You will receive them sometime this week. Your only task now is to provide an adequate English translation of the Japanese global Design Directory for 2012. We will start from the regional office in London in April. You have almost three weeks. That should be enough time."

LONDON PRESENTATION

Yukiko set to work. She knew it would not be easy to persuade the British team to give up a big part of their freedom. And working on the presentation for the trip, she found another issue that could make her task even more challenging. The Japanese design and marketing teams provided her with a number of Japanese text slides on the proposed Design Directory 2012.

The slides had been approved by Mr. Hirano. They included elaborate text descriptions of the major elements of the product image and addressed the main features of the global Design Directory for the year 2012. A few slides of the presentation talked about the present-day global situation, the consequences of the world financial crisis and the lack of stability and hope. Because of this, the presentation suggested that the 2012 design should reflect a longing for hope, joy, stability and comfort in order to bring some optimism to their customers. As many as three slides were also used to emphasise the importance of copyright and disclosure requirements as well as possible consequences of a copyright violation. "How much trust does headquarters have in our subsidiaries?" she asked her colleagues. "Mr. Hirano's wife's bag was stolen in Madrid airport. Since then, he does not trust anybody who looks Spanish or speaks Spanish," they told Yukiko.

The translation work proved to be a much bigger challenge than Yukiko expected. The Japanese abstract's intricate and formal style was difficult to put into English. She could not find adequate English equivalents that would sound as beautiful as in her native language and at the same time would perfectly reflect the meaning of each idea. Yukiko spent hours searching for synonyms and possible translations. Her foreign colleagues did their best to help, which was not much, as most of them could manage only spoken but not formal written Japanese. Mr. Hirano was of almost no help and Yukiko did not want to challenge his English skills either.

Yukiko did her best to find seemingly fitting words and phrases: comfort; vintage and tradition; cosiness and warmness; new geometric thinking with straight lines and simple structures; rediscovery of nature as well as multitude and diversity of materials; active cool with the features of functionality; performance and new possibilities (Figure 4.7).

Main concept of the collection for 2012 of GDG.
Longing for safety and new spirit of simple forms

As a consequence of economic crisis, people will long for joy and
the recovering of optimism needs to be supported of
traditional values and gentle feelings, by means of
combination of the elements of harmony, nature, comfort,
vintage, optimism, active simplicity, new possibilities, new
cool geometric.

The new design brings comfort to the shaken by the
crisis society by means of simple, natural textures and
imitating nature materials and simplicity, combined
with the bright, nature related colours.

Figure 4.7   Global Design Directory sample slide 1. In order to make her message and ideas
as unambiguous as possible, Yukiko provides detailed descriptions rather than only images.

"Active cool," for example, is the way Yukiko managed to translate the original
Japanese idea (a combination of kanji signs). One slide of Japanese text often turned
into two slides of English text. Yukiko also included as many images as she could,
hoping they would help people to better understand the meaning of the text (Figures
4.8–4.9). The final presentation that was supposed to last for a maximum of 30 minutes included more than 40 slides.

Combinations/Fusions

- Inspired by nature the
  combination of opposites,
  natural elements, water and
  sun, colours, blue and red.
- Handy craft fusions.
- Unusual elements for the
  visually challenging
  arrangements make attractive
  look.
- Natural colors offer up new
  visual aspects.

Natural Look

- Dull and original look of traditional
  textures.
- Flowing structures offer up harmony
  and continuum.
- New perceptions of natural flowing
  surfaces celebrates new values.
- New space feel of nature invites a
  new level of surface consideration

Neo Geometric
Reflects gentle impression

Figure 4.8   Global Design Directory sample slide 2. Extra images are expected to increase
understanding of the design concepts.

Key colors

- Red – sun. New era color emerges. Sun as symbol of new rise, history, old values, traditions, nature, universe reflects natural look.
- Blue – sea, cold metal, bamboo. Natural blue leads universal color. Comes from natural universe-reflecting colors of blue water, warm sky, gentle greenery.
- Sophisticated contrast of colors celebrates a clear definition of warmth and hope.
- Main white background celebrates clear and contrast presentation.
- Elegant silver border emphasizes natural resources and elegance.

Figure 4.9 Global Design Directory sample slide 3. Colours are painstakingly described. In Japanese, the word "blue" can be used to describe certain green-looking objects. This leads to further confusion when the slide is translated into English.

Yukiko tried to practice the presentation and found that she had to "chatter like an insane summer cicada" to say everything she had to. She complained to her American friend that in spite of the general image of the company as a "modern and democratic organisation," she did not have enough freedom to make major changes to the Design Directory slides provided by the marketing and design teams as they had already been approved by her boss. She managed to get a few proposals through and finally reduced the number of slides to 37, cutting off one of the copyright violation warnings.

The day of the England trip came too fast. The way to London took much more time than was planned because of the unexpected eruption of Eyjafjallajökull, the volcano in Iceland. As a result, the plane had to land in Paris instead of London. Yukiko spent a few hours trying to arrange train tickets for the Japanese team that included Yukiko's boss, Mr. Hirano, and the head of global marketing, Mr. Matsuo. By the end of the day she could even pronounce the volcano's name without stumbling. They arrived in London 18 hours late. Everything went wrong from the beginning. Yukiko thought it was a bad omen.

Simon McMahan, the leader of the British design team, met them at the airport. Simon joined Green Hills in 2008 as a designer but was quickly promoted to design group leader. Yukiko had known Simon when she was studying in London. He was a creative, talented, thinking-out-of-the-box designer. Yukiko liked his "very English style and the inimitable British pronunciation." Mr. Hirano himself agreed that Simon was surely reasonable, well brought up and not as noisy as his American colleagues. Simon had had no experience working in an "Asian environment" and once experienced serious difficulties communicating to his Japanese colleagues. "It doesn't matter how much I try to be understood, it simply seems like they cannot give a simple, clear reply to a simple, clear question. It's like I'm playing some strange 'hide-and-seek' game with the Japanese team," he would complain to his friend. Apart from this, he liked his job and all the freedom the Japanese gave to his design team.

The Japanese team had much work to do the next day. Except for the meeting with the designers, Mr. Matsuo and Mr. Hirano were planning to meet with the marketing and product development groups. On the way to the office, Simon proposed to change the schedule, to skip some formalities and get down to business. Mr. Hirano thought the proposal over and told Yukiko to shorten her presentation to 15 minutes. "Oh, please, how?" thought Yukiko, "How am I expected to say everything in such a short period of time?" But there was no sense in arguing and she decided to speed up, skip a couple of slides and hope that the English team would understand everything even if she shortened her talk.

The conference room was prepared and the English team was already waiting for their arrival. Mr. Hirano made a short introduction first. "We have some changes coming in the design procedures," he said. "Ms. Omura will share with you all the details. She will also introduce you to the requirements of the new Design Directory for the year 2012. Thank you."

Yukiko felt her nerves were betraying her. Although she had originally planned to explain the new policy in her own words, she felt that she started speeding up, and in order not to get confused, she simply started reading the information from the slides. She spent 10 minutes talking about the new policies. Her team had spent seven months working on the concept of "longing for safety and new spirit of simple forms." She had to quickly explain that the new design would bring comfort to a crisis-shaken society by means of simple, natural textures, materials that imitate nature and simplicity combined with bright, nature-inspired colours. It was Yukiko's favourite part and it was disappointing to spend so little time on this topic.

The British team looked interested. To them, the Japanese presentation looked like a kimono or an Asian painting, detailed, ornamented, elaborate and consisting of many parts, layers, elements and small details. Each slide offered a multitude of components: text windows, images, schemes, tables and graphs. At times it was difficult to decode their meanings and purposes. By the end of the presentation, most of the questions were circulating around the unification of the global brand. The British team was interested in the reasons for the design unification, how the new policy would affect their work, if any layoffs are planned, and so on. Yukiko and her group leaders were surprised that, in spite of the detailed slides, they seemed to be explaining the new design policy all over again. Yukiko also felt that the British team did not show much loyalty. Their questions were frank and unequivocal, and the discussion was intense. Although Mr. Hirano and Mr. Matsuo were adept at reading and writing English, spoken communication was difficult, so Yukiko was doing much of the talking. Yukiko felt exhausted by the end of the presentation but was relieved that at least the new Design Directory seemed to be understood and accepted.

In July, the British team was expected to visit the headquarters and present their design ideas for the year 2012. In spite of the complicated and chaotic trip back to Japan, Yukiko could finally relax and forget about the old-fashioned Japanese superstitions that a bad start of any undertaking is a bad omen.

OSAKA "DISASTER"

The British team's July visit and presentation made Yukiko think that her fear of bad omens was not groundless after all and that she should visit a temple and then start looking for a new job.

Simon came alone. He offered a 12-minute colourful presentation that included six slides with beautiful, big images and a couple of key words on each slide, such as "Real Me," "Rejoice," and "Back to the Roots" (see Figures 4.10 and 4.11).

Simon was confident and friendly. Two company translators provided word-to-word translation of everything he said. Yukiko quickly noticed that only a few of the original requirements were considered for the design. It was a free interpretation of the original concept. The British designers appeared to have either misunderstood or intentionally changed the concept. The slides showed images of hippy style, well-worn, somewhat shabby boots, fire, simple colours and flowery design elements. The key phrases reflected natural life, individuality, and personal freedom. Only the idea of unification with nature resembled the proposal of the global design group.

Simon, at the same time, was uncomfortably surprised to see that in spite of the great importance of the visual elements of the presentation, a big part of the Japanese team seemed to be sleeping. They closed their eyes and showed some signs of attention only when the interpreter provided the translation for the slide. After that they fell asleep again. It wasn't motivating, but he had no choice and tried to address those few people in the room who looked interested and even nodded from time to time. By the

Figure 4.10 British team presentation slide sample 1. Simon is convinced that designers worldwide speak the same visual language and can understand each other without too many words.

Figure 4.11 British team presentation slide sample 2. The British presentation is based on image, impressions and feelings.

end of the talk, almost no questions were asked and no ideas discussed. Mr. Hirano thanked Simon for his "exciting" presentation and suggested going for lunch.

Before lunch, Mr. Hirano pulled Yukiko aside. "Omura-san, you saw the presentation. What do you think?" Yukiko did not know how to reply. She had to agree that it did not meet the requirements of the Design Directory in the most part. "This is an unpleasant issue, Ms. Omura. Please, convey to Mr. McMahan that their ideas would require some additional work. The word 'back' implies no progress. It rather signifies regress. 'Hippy' is somewhat old-fashioned and does not match the original idea of simplicity. 'The real me' is too individualistic. I believe you can also see the obvious misinterpretations of the original concepts. Do you think your English colleagues would make such a big mistake or could it be deliberate? You understand we have a problem here, don't you?"

"Hirano-sensei, I am sure they simply made a mistake. I will talk to Mr. McMahan and explain the situation," Yukiko said.

Nothing was said during the lunch. Mr. Hirano and Simon had a pleasant chat about the World Cup and the performance of the national British and English teams. Before the official dinner, Yukiko came to Simon. "Simon, I am afraid we have some difficulties with your proposal. Do you think you could come up with another design proposal by the end of the month?" Simon was taken aback and first did not know what to say. He simply gave Yukiko a puzzled look and kept silent. He did not expect such an outcome and needed some time to think it over. Most of the Japanese, even those who slept during the talk, Simon thought, looked satisfied by the end of the presentation.

At the dinner, neither the Japanese nor Simon was relaxed. Yukiko felt she should either find out what went wrong and offer solutions before the same story happens with another subsidiary or probably better try her luck at another company. Months of work were wasted. Her career was jeopardised. She just wanted to go to bed and forget the whole thing. Besides, she felt sorry for Simon and did not feel confident to come and talk to him. She was sitting at her table, almost did not touch her food and felt so possessed by these thoughts that she did not see Simon coming to her.

The voice of Simon abruptly brought her back to reality: "Listen, Yukiko. We cannot afford to foul this up. What if something goes wrong!? I am afraid you are the only person who could explain to me what is going on. I made a few attempts to talk to Mr. Hirano, but he keeps on changing the topic or assures me that everything is fine and all the small issues would surely be solved. I feel mistrusted, not taken seriously and completely confused. I need your help." Yukiko felt so relieved and grateful to Simon that he started this conversation. "There is an empty table near the bar, should we move there and talk?" suggested Yukiko. They took their glasses and went to the relatively quiet corner of the restaurant.

"I'm completely lost, Yukiko," Simon said. "Why would they shoot down our design proposals? I thought we did a great job following the global design!"

Yukiko answered, "I understand that you are very upset, Simon. But we have a few issues with your proposal. It does not meet the requirements of the original Design Directory. I do not know how this could happen. You did not ask any questions after my presentation in London. I was sure you understood our intentions and ideas."

Simon then said, "Yukiko, I have to confess, we did have some problems with your proposal. In fact, we were so frustrated with the first part of the presentation that we did not pay much attention to the second part. There was so much background information that we were lost on the fifth slide. The policy was detailed, but did not take into account our main concerns. You very nicely told us about the step-by-step procedures, our tasks and the company's concerns but didn't address our interests. We felt we were instructed like small, unreasonable kids. You offered us no choice. It looked rather like a lecture."

Yukiko was shocked. She never thought the first part of the presentation could be seen this way.

Simon continued, "We were so worried about all the changes coming and the time was so short that we had no time to talk about the design directory properly. We decided to rely on the images you provided as we did not understand the text descriptions on the slides. They were a bit too vague."

Yukiko replied, "Simon, I am sorry, I really did not mean to behave like an arrogant teacher. I probably relied too much on the standard Japanese communication style, trying to appeal to your loyalty and trust in the company. Naturally, we had no intentions to make your life more difficult and no lay-offs were planned. We only wanted to change the procedures…but probably did not find the right way to deliver our intentions to you. To be honest, Mr. Hirano and Mr. Matsuo were quite unhappy with your aggressive questions after the presentation, too."

"Were they? Trust me," Simon said, "we did not mean to be offensive. We only used the opportunity to air our opinion. Do you think we could talk about my disastrous presentation?"

"Sure," agreed Yukiko, "you don't have a lot of time to come up with an alternative proposal."

"All right then," Simon said, "let us start from the very beginning. Why are they unhappy? They looked quite satisfied! Nobody questioned me by the end of the talk."

"Simon, I think this time you lost them on the second slide," Yukiko replied with a smile. "You gave no information at all: only impressions and emotions. Sure, they listened to you, sure they nodded their heads. Nobody wanted to offend you. But, in fact, your English was simply too complicated and the translators gave a word-to-word translation of all the idioms you used. Even I could not figure out what exactly you wanted to say from time to time."

"But Yukiko, they didn't ask questions," remarked Simon.

Yukiko said, "They didn't know what to ask and they didn't want to offend you in front of other people. I bet you will learn a lot more tomorrow, when talking to them privately and face to face. Just listen carefully and do not expect them to be as direct as you might be."

"Now, what about my slides?" asked Simon.

Yukiko answered, "They were too abstract. You gave us no information. We could not see your logic. Give some more details next time. Text descriptions would also be very helpful. Do you have your laptop? I would like to explain to you once again the main ideas of the Design Directory for 2012."

"Good idea," Simon said. "Just give me a minute to start the computer. Tell me, why did they sleep through the presentation?"

Yukiko laughed and said, "Simon, they weren't sleeping. Most likely, you simply overloaded them with your bright and abstract visuals and they made an attempt to concentrate on the translations provided."

For the next two hours, Yukiko explained to Simon the Design Directory 2012 in detail and, wishing him a good night, rushed to catch the last train back home. It had probably been the longest and most informative dinner in Yukiko's career. It convinced her that the whole project was on the verge of failure because of the strong assumptions held by those who were involved in it. It was naïve and dangerous for anyone to assume that other people see the world the same way or think and work the same way. The Japanese and British teams tried to use the same language to communicate, but they operated from two different worldviews and sets of values. The indirect, relationship-oriented and disciplined Japanese way of doing business was misinterpreted by the more direct and task-oriented British team. Yukiko learned the hard but valuable lesson that she needed to be careful with assumptions and to adjust to the style of her international colleagues. "When in London," she thought to herself, "do as the Londoners do."

JOURNEY AHEAD

The next morning, after a very short night, Yukiko came to her office unusually early. She needed to talk to Mr. Hirano and she was quite sure he would want to listen to her report as well. Mr. Hirano summoned her to his office at 9:00 and said: "I saw you talking to Mr. McMahan yesterday night. I believe you both had a lot to discuss. Could you explain to him all the issues?"

Yukiko knew this conversation with Mr. Hirano would not be easy. Her boss surely did not need excuses and explanations. It was her task to solve the problem. This is why she chose to simply say, "Mr. Hirano, you are absolutely right. We did have a lot to discuss in order to improve our performance. Mr. McMahan assured me that no disrespect or offence was meant. The British team simply misunderstood some of our ideas. Mr. McMahan will take care of these issues and will come up with new design suggestions. He was also kind enough to point out some weak points of my presentation. There are some elements that require improvement. I will take care of them immediately."

Mr. Hirano listened to her carefully. He nodded a few times and said, "I see, Ms. Omura, it is good you could explain everything to Mr. McMahan. He is a good designer and I am sure he will come up with a good suggestion soon. You, Ms. Omura, please consider the criticisms of Mr. McMahan, work on the presentation and show me the improved version sometime next week and work harder."

The next presentation was supposed to take place in Sidney. Yukiko hoped that Simon's advice would help her avoid major misunderstandings with the Australian team. At least, hopefully, there would not be another natural disaster to make matters worse.

RECOMMENDED READINGS

1. This story suggests important differences in Japanese and British presentation styles. Other scholars also argue that cultural differences lead to different conventions between Japanese and English communication, in both expository writing and technical communication. On the other hand, some scholars believe that these purported differences may be superficial, exaggerated, or even misunderstood. The following sources provide information on these beliefs:

   - R. Kaplan, "Cultural thought patterns in inter-cultural education," *Language Learning,* vol. 16, no. 1–2, pp. 1–20, 1966.
   - W. Fukuoka and J. H. Spyridakis, "The organization of Japanese expository passages," *IEEE Transactions on Professional Communication,* vol. 42, no. 3, pp. 166–174, 1999.
   - R. Kubota, "A reevaluation of the uniqueness of Japanese written discourse: Implications for contrastive rhetoric," *Written Communication,* vol. 14, no. 4, pp. 460–480, 1997.
   - D. Cahill, "The myth of the 'turn' in contrastive rhetoric," *Written Communication,* vol. 20, no. 2, pp. 170–194, 2003.
   - W. Fukuoka, Y. Kojima, and J. H. Spyridakis, "Illustrations in user manuals: Preference and effectiveness with Japanese and American readers," *Technical Communication,* vol. 46, no. 2, pp. 167–176, May 1999.
   - J. H. Melton, "Going global: A Case study of rhetorical invention, packaging, delivery, and feedback collection," *IEEE Transactions on Professional Communication,* vol. 52, no. 3, pp. 229–242, 2009.

- K. Maitra and D. Goswami, "Responses of American readers to visual aspects of a mid-sized Japanese company's annual report: A case study," *IEEE Transactions on Professional Communication,* vol. 38, no. 4, pp. 197–203, 1995.
- D. L. Major and A. Yoshida, "Crossing national and corporate cultures: Stages in localizing a pre-production meeting report," *Journal of Technical Writing and Communication,* vol. 37, no. 2, pp. 167–181, 2007.

2. This story describes the difficulty of translation between Japanese and English. The following sources provide more information on this topic:

- M. Burnette, "Managing English-to-Japanese translation projects," *Technical Communication,* vol. 39, no. 3, pp. 438–442, 1992.
- Y. Furuno, "Translationese in Japan," in *Translation and Cultural Change: Studies in History, Norms and Image-Projection,* E. Hung, Ed. Amsterdam & Philadelphia: John Benjamins, 2005, pp. 147–160.

3. To learn more about the Japanese culture, consider the following sources:

- E. T. Edward and M. R. Hall, *Hidden Differences: Doing Business with the Japanese.* New York: Doubleday, 1990.
- M. L. Maynard and S. K. Maynard, *101 Japanese Idioms: Understanding Japanese Language and Culture through Popular Phrases.* Lincolnwood, IL: NTC Publishing, 1993.
- R. J. Davies and O. Ikeno, *The Japanese Mind: Understanding Contemporary Japanese Culture.* North Clarendon, VT: Tuttle Publishing, 2002.
- S. Traweek, "Border crossings: Narrative strategies in science studies and among physicists in Tsukuba Science City, Japan," in *Science as Practice and Culture,* A. Pickering, Ed. Chicago: University of Chicago Press, 1992, pp. 429–465.

DISCUSSION QUESTIONS

1. Several years of studying in the United States and England had had a certain effect on Yukiko. Try to picture what these effects might have been. For example, which of Yukiko's actions and comments seem to be affected by her knowledge of English culture? Which ones are in keeping with her native Japanese culture?

2. Depending on the workplace culture you are familiar with, some of the Japanese workplace practices described in the story may or may not strike you as surprising. As Yukiko put it, she was doing two jobs with the pay of one, had to stay late in the office, and often worked from home. Do more research on the Japanese workplace protocol and compare it with what is commonly practiced in your region or culture.

3. The word "loyalty" was mentioned multiple times in the story. Research what

"loyalty" means in the Japanese workplace and culture and discuss its significance in the story.

4. "Personal differences" was not a favored term for Mr. Hirano, but what role would you say personal differences have in cross-cultural communication?

5. What do you think of Yukiko's translation of the Japanese presentation slides? What do you think of the English diction she used? If you are bilingual, what might you do in a similar situation where you have to translate something that is seemingly untranslatable?

6. Given Yukiko and Simon's discussion, how might your revise their two presentations so they can be more effective with their respective audiences? Or do you think much more should be done than redoing the slides to help prevent similar miscommunications in the future?

CHAPTER 5

COLLECTIVE LEARNING IN EAST AFRICA: BUILDING AND TRANSFERRING TECHNICAL KNOWLEDGE IN LIVESTOCK PRODUCTION

K. Alex Ilyasova and Cheryl Birkelo

K. Alex Ilyasova is an assistant professor and director of the Professional and Technical Writing Program at the University of Colorado at Colorado Springs. Her research interests include writing program administration, with a focus on technical communication programs, and identity and literacy studies. She has most recently published an editorial in Programmatic Perspectives (March 2012), "An Editorial Introduction of a Curriculum Showcase." Additionally, she has a forthcoming co-authored book chapter on the experience of working as the lone faculty/administrator of professional, technical, and scientific communication programs. Her previous publications include book chapters on the issue of visibility and identity in LGBT studies and queer business practices.

Cheryl Birkelo is an instructor in the Professional and Technical Writing Program at the University of Colorado at Colorado Springs. Born and raised on a working cattle ranch and Grade A dairy farm, Birkelo was a volunteer 4-H leader/agriculture extension program educator and a 10-year 4-H member. Prior to starting at the University of Colorado at Colorado Springs, Birkelo was an instructor at the U.S. Air Force Academy and a technical writer for a private engineering firm in Woodland Park, Colorado. Before returning to Colorado, she was a technical writer, lab-field technician for South Dakota State Wildlife Fisheries Unit. Birkelo's special areas of interest for study are rhetoric of science and argument, American nature

Negotiating Cultural Encounters. Edited by Han Yu and Gerald Savage
Copyright © 2013 The Institute of Electrical and Electronics Engineers, Inc.

103

writing, and eco-feminism. She recently presented a paper at the Society for the Study of Midwestern Literature, "Gene Stratton Porter: Scholar of the Natural World," at Michigan State University.

CHAPTER SYNOPSIS

With the assistance of project staff, Ilyasova and Birkelo describe in this story a large, non-governmental organization (NGO) development project in East Africa. The project provides the local agricultural extension programs and smallholder farmers with techniques, trainings, and resources to enhance their livestock production. Paul, the project manager, is a U.S. livestock nutritionist who had years of academic and industry experience and previous development project experience worldwide. This was, however, his first project in the East African regions. The main project staff included Petros and Tesfaye, two local technical advisors who were familiar with the regions and were fluent in the local languages, even though they lacked experience in some of the technologies involved and their applications. The three worked together, through both informal and formal learning sessions, to understand each others' knowledge and skills, to design ways to transfer these knowledges and skills to the local audiences, and to develop targeted training materials and methods. As the project progressed through its multiple stages, the team faced the challenges of bridging various kinds of cultural, linguistic, educational, economic, and technological "chasms."

East Africa and the Horn of Africa are lands of extreme geographical and environmental diversity: from mountain ranges to semideserts, from rivers with hippos and crocodiles to sandy, volcanic rock-strewn expanses with little to no water, and from lush semitropical highlands to acacia-treed, grassy savannahs, all environments which pose extreme challenges for the peoples attempting to farm and survive there.

Livestock production is a key facet of East African agriculture. In the poorest countries, it contributes as much as 30 to 35 percent of agricultural gross domestic product (GDP) and 15 to 18 percent of export income while directly employing a third or more of the rural population. In some regions, if we consider related industries (e.g., leather processing for shoes) and the widespread use of livestock for draft and transportation, over 80 percent of the population, even those in urban areas, are dependent to some degree on livestock production. In light of these statistics, it is no surprise that improvements in livestock production have the potential to contribute significantly to the economic and food security of East Africa.

Although a growing number of Africans are well-educated, capable professionals with expertise in livestock nutrition and management, they are nonetheless short in number relative to the need. Additionally, in an effort to meet the food needs of their growing populations, often through the introduction of increasingly sophisticated technology, African countries continue to see gaps in expertise. To fill such gaps, some governments work with nonprofit, nongovernmental organizations (NGOs) that can provide technical expertise—usually based on western scientific practices but seeking ways, at least in the best instances, to work with local conditions and local knowledge, and to partner with native-born and educated specialists.

LIVESTOCK PRODUCTION PROJECT

One of these NGO projects focuses on livestock production in East Africa, with the overall goal to improve farmers' livelihoods and increase their income through improved livestock productivity and efficiency. This large, multistage project provides training, technical assistance, and material support to help livestock owners enhance their skills and capacities producing and managing animal feeds and forages. To reach its goal, the project must rely on agricultural co-operatives and agriculture extension programs. Agricultural co-operatives provide farmers with a method to pool production and/or resources (seeds, herd stock, machinery, technology, etc.) and to market their various products. Agriculture extension programs are supported by the local government and modeled after the U.S. Department of Agriculture Cooperative Extension Service land grant university programs. They strive to bring training, technical assistance, and material knowledge to the local level. The co-operative farmer leaders (or co-op leaders) and agriculture extension agents (or ag agents) are key members to help the livestock development project reach individual farmers.

In the fall of 2009, the NGO hired a new project manager, Paul, a livestock nutritionist from the United States. (The names and identifying characteristics of the people and regions mentioned in this story have been changed or omitted to maintain confidentiality.) Paul's background spans both academia and production agriculture. He has graduate degrees in animal science, specifically ruminant nutrition and management, and has spent half of his 30-year career as a university teacher and researcher in the United States. The other half has been spent in commercial livestock production and technical support to allied industries, such as commercial feed manufacturing.

Paul has always felt that his practical, applied background was just as important as his academic qualifications; the former informs the latter and vice versa. He can discuss topics ranging from laboratory techniques for determining nutrient content of feeds to how to take a hammermill apart and put it back together. When asked, he usually says, "I grew up with burn holes in my pants, busted knuckles, and grease under my finger nails," referring to the many hours spent over the years working on vehicles and farm equipment with acetylene torches, arc welders, and wrenches.

Nevertheless, on the surface, Paul—a middle-aged white male—might seem out of place in the East African landscape and culture. When an NGO staff member expressed concern during his interview that he might be too academically oriented, Paul's response was "I still have shit on my boots." More practically, he could also point to the fact that he was raised, in part, in West Africa where his father had worked on development projects. Paul has also worked on short-term projects in Latin America, other regions in Africa, and central Asia. Despite, or because of, this background, Paul knew transferring the knowledge and experience he possessed to the project's target beneficiaries would be neither easy nor simple. Cultural, linguistic, educational, economic, and technological "chasms" would need to be bridged.

Paul got the job and now lives in the East African highlands 11 months out of the year. As the project manager, Paul's first job was to get to know the staff. The project had started earlier in 2009 with a different manager, so the staff was already in place when Paul joined the project. As project manager, Paul worked most closely with Petros and Tesfaye, two local technical advisors in the project head office.

Petros is a young man in his early 30s. A native of East Africa, he received his bachelor's degree from a regional agricultural university. Subsequently he studied in Europe, ultimately earning a Ph.D. in animal nutrition. He is academically strong but somewhat short on practical, applied experience. However, he is quick to grasp new knowledge and skills. It usually takes only one explanation or demonstration for him to catch on. Just as important, he is very fluent not only in the common language of the region but in English as well. His writing skills are also well developed. This is undoubtedly due not only to his own intellectual abilities but also to the fact that he was raised in a home where education was considered most important—Petros's father is a teacher at a local elementary school.

Tesfaye is a local in his mid-50s with a master's degree in agriculture from a regional university. His educational background and grasp of the science of livestock nutrition and management are not as strong as that of Petros'. His command of the English language and writing skills are not as well developed either. However, Tesfaye has many years of practical experience, having worked in livestock production as well as for the Ministry of Agriculture in the area of livestock marketing. His understanding of local livestock production and trading practices commonly used in this part of Africa helps keep the project grounded. He is very good at relating to ag agents and farmers in the field. Because of his years in government service, Tesfaye is also good at tracking down information, resources, and outside expertise that are often needed for the preparation of training materials.

Together, Paul, Petros, and Tesfaye embarked on a four-stage process—which is typical for projects like this one—in the development and the subsequent use of training techniques and support materials.

Stage 1 involves identifying the specific knowledge and skills of the ag agents, co-op leaders, and smallholder farmers so the project staff could design teaching approaches and training materials that are tailored to their needs.

The second stage is often referred to as the "training of trainers," or TOT in project shorthand. It is not possible for the small group of project staff to directly train the thousands of co-op staff and smallholder farmers scattered across the various regions. However, it is possible to assist the ag agents in upgrading their own knowledge base and skills so they can, in turn, train co-op staff and farmers in their own locales. This way, TOT creates a multiplier effect in the dissemination of knowledge and skills.

At the third stage, training materials are adapted to the local Afro-Asiatic languages, knowledge, and skill levels, and the ag agents give subsequent training to co-op staff and farmers. Such outreach and access to the co-op staff and individual smallholder farmers are primarily done through livestock and multipurpose co-op agricultural unions located throughout the region. The co-op unions help individual farmers group together to gain better market access and bargaining power, to procure modern materials and services, and to promote education and training. These unions also provide a location and focal point for specific training activities.

The last stage involves project site visits to assess whether the project has had its intended effect. Have farmer knowledge and skill sets been improved and do these translate into greater productivity and income? Lessons learned from such visits are used to adjust approaches where necessary for future projects.

DEVELOPING TRAINING MATERIALS

Since Paul joined the project midway, one of his first challenges was figuring out exactly how much Petros and Tesfaye knew about the various subject matters for which they would later be producing training materials. Paul reviewed their employment files, which included their curriculum vitae and answers to job interview questions. However, such documents provided only clues. To get on the same page, Paul gathered this small group together to collectively work out the main topics and subtopics to be covered in a particular training documentation: a feed manufacturing brochure and manual. The brochure would be used as a handout in the local language/dialects for individual farmers and as a reference after a training workshop. The manual would be a more complex document for the ag agents and co-op leaders to use in producing feeds and rations at feed milling sites.

Paul started the session by providing information based on his experience in the U.S. and what he felt should be included in the training documentation: feedlot, dairy nutrition and feeding management, forage production and pastureland development, feed formulation and manufacturing, and study tours of feed mills and livestock operations. Whether African or North American, farms tend to have universal elements with specific regional requirements. Paul could provide the universal practices, and Petros and Tesfaye were encouraged to "fill in the details" by contributing their own ideas, particularly with respect to making it fit into the local context.

As the work sessions began, Tesfaye's history of working with local livestock producers, especially in transitioning farmers from free-range animal production to confinement operations such as milking barns and feedlots, provided the firsthand knowledge they needed.

Petros also chimed in. When the team was discussing how to produce molasses lick blocks—which are used to supplement ruminant livestock with key nutrients missing in their normal diets of straw and other high-fiber crop residues—he commented that the blocks should be adapted for specific types of livestock: "In this region, sorghum forage is commonly used; not teff straw. It's a cheaper source of feedstuffs so we can produce least cost rations with readily available products for the farmers. We'll see more camels here than in the sugar plantation region where there are more dairy cattle, too."

"Well then," Paul replied, "we'll have to adapt the lick block nutrients for camels as well as cattle. Where will the camels be penned, at the same facility with the cattle or somewhere else? What's the base price for urea there, Tesfaye? We have to keep the lick blocks affordable for the livestock producers."

"Camels will be in separate pens but at the same feedlot," Tesfaye responded. "I'll check with the local distributor at the co-op for the current price in this region. Their real problem is access to enough water supplies. The government water line project hasn't reached here yet and they're having to haul water. That's really costly and hurts their profit margin."

"How are they dealing with those costs? Are they adding it to the production costs before or after they sell and ship the stock?" Asked Paul.

"That'll depend on the individual contracts the feeder has with the customer, so it'll vary as to how they charge or absorb those costs," Tesfaye answered, while Petros nodded his head in agreement.

During these sessions, Tesfaye and Petros also helped to identify the specific knowledge and skills that the ag agents, co-op staff, and finally, smallholder farmers would likely have.

A collective work dynamic developed between Paul, Tesfaye, and Petros during these first few brainstorming sessions. With this dynamic seemingly in place, everyone was encouraged that each person would be easily responsible for developing a first draft of his assigned topics for the brochure and manual. Paul then took on the responsibility of pulling together and editing the various pieces of writing produced.

However, shortly after receiving the first drafts, Paul realized that they had more challenges still. As trainers, Petros and Tesfaye would be providing applied training sessions to ag agents and other government officials (project step 2) who then, in turn, would train individual farm families (project step 3). Therefore, it was important that Petros and Tesfaye not only had the conceptual knowledge but the applied knowledge and know-how to do the things they included in the manual and brochure. Although their writings were conceptually correct and contained useful information, they were missing the practical information that comes from knowing how to carry through a procedure.

For example, Figure 5.1 gives the section of "mixing" in the manual written by Petros and Tesfaye.

3.1.3. Mixing. The objective of mixing is to create a homogenous blend of multiple feed ingredients. Each bite of the resulting blend should provide the same nutrition as the last. Needless to say, the simplest tool for mixing is the shovel. A shovel is perfectly adequate for mixing small quantities of feed provided the person carrying out the task pays attention and does a thorough job. For larger quantities several types of mixers are commonly used. Each has its strengths and weaknesses. Some mixer types are better suited to certain production situations than others. The most common mixer types are....

Figure 5.1 Manual section on "mixing." This original section contains correct and useful information but lacks practical information that comes from knowing how to carry through a procedure.

Descriptions and photos of the main types of mixers (i.e., vertical, horizontal, and drum) followed the section.

Pointing to this section of the draft, Paul asked Petros and Tesfaye, "Do you know how to do the mixing?"

"Yes," both of them answered quickly.

Thinking for a moment, Paul asked again, "Have you done mixing before?"

Petros and Tesfaye looked at each other tentatively and both said, "No."

Paul realized that their answer to the first question was based on the fact that they were well familiar with procedures devised and published by others and thought that based on this, they should be able to accomplish the task. But they had not done some of these procedures themselves. Paul worried that such "recipe" understanding could miss practical details that can trip one up when trying to demonstrate the procedures for others. To gain this practical knowledge, Petros and Tesfaye performed the tasks themselves. Moreover, to sustain these practices beyond Petros and Tesfaye—because individuals come and go in the local communities—Paul modeled an informal learning and sharing structure.

An example of this informal learning and sharing involved the making of the molasses-based lick blocks mentioned earlier. Paul, Petros, and Tesfaye each contributed to the writing of the instructions, which were printed on 9 × 11 sheets of plain white copy paper. Tesfaye and Petros performed a usability test of their written instructions to make sure they can subsequently share this knowledge with others.

On the date of the test, Petros and Tesfaye gathered at Paul's home in the carport with all of the supplies necessary to complete the task. They went through the four basic steps: (1) weighing ingredients, (2) mixing ingredients, (3) pressing the mix into a mold to compact and create the desired shape, and (4) curing (drying and hardening). Paul stood back as an observer and coach as Petros and Tesfaye read to one another and performed the steps. Petros and Tesfaye had a typed "recipe" sheet (Figure 5.2) and began by converting the ingredients from percentages to units of measurement for either dry matter or liquids.

Tesfaye's years of applied experience showed, as he was more confident in his actions. Petros was adept at doing the math in his head but a bit awkward in doing the mixing. They did a verbal back-and-forth exchange throughout the process, dou-

Molasses	31%
Wheat bran	25%
Oilseed meal	13%
Cement	15%
Urea	10%
Salt	3%
Trace mineral	3%

Figure 5.2 Molasses-based lick block recipe sheet. This sheet was used to calculate ingredients in the creation of molasses-based lick blocks.

ble-checking one another, their figures and quantities, and ticking off completed actions. Meanwhile, the "recipe" sheet had curled and crinkled and the ink ran since it was a drizzly rainy day. They quickly realized that a laminated instruction sheet would be a helpful addition for the farmer as environmental conditions for doing the process will differ and paper will easily disintegrate.

Petros and Tesfaye did not have a cement mixer or any machine mixers so they did the work manually with buckets, tubs, plastic pitchers, and empty plastic water bottles—all materials that the common farmer would have at hand. They suspended a scale from the rafters of the carport to weigh the materials before adding it to a mixing bucket. The dry ingredients were combined first and Tesfaye stirred while Petros went to the water bucket to fill a two-liter plastic bottle.

Before Petros upended the bottle, Paul offered a word of advice: "Watch how much liquid you add, Petros. You only need enough to moisten the mix. The molasses adds more moisture than you think."

"OK, say stop when it feels right, Tesfaye," Petros said.

"That's good. Whoa!" Tesfaye replied as Petros poured a steady trickle.

Once the cement was added, it became a challenge to mix the ingredients and they scrambled to find a stick or piece of rebar that would hold up to the heavy consistency.

"Today's rain isn't going to help us," Paul said, as he looked up at the dripping and leaking carport. "Blocks will take longer to dry and cure."

Petros stood back and tried to strike a match while his cigarette dangled over the dry mix. The guys joked that his cigarette ashes weren't the desired additives in the mix. He quipped back that cigarette ashes blended into the cement just fine.

Finally, they had it all mixed and test blocks were molded with a shoe box, which farmers would have, as well as with a block press (see Figure 5.3), which co-opera-

Figure 5.3   Block press. A block press is used to create molasses blocks.

tives might use if they are making these blocks in large numbers for sale. The molasses blocks then sat out on the carport for several days curing.

During this and other hands-on workshops and testing, a feeling of community developed among Paul, Petros, and Tesfaye. Their informal collaboration provided a chance for them to talk about what they were doing on this project and to establish a sense of trust and sharing. These workshops also helped Petros and Tesfaye to add practical tips to their manual (see Figure 5.4).

TRAINING THE TRAINERS

Once Petros and Tesfaye felt they had the practical experience with the various techniques, they moved on to stage 2 of the project: training the trainers and the ag agents. At the beginning of the development process for each training segment, a great deal of discussion went into how advanced to make the training materials. Although ag agents were all required to have at least a bachelor's degree in agriculture, they were nonetheless a very diverse group because of the quality of the school they went to, their area of specialization within agriculture, their command of the common regional language, and their years of experience in their jobs.

Tesfaye was concerned that they not shoot too high. "Some of the participants," he cautioned, "may not understand such things." But Petros came back with, "They all have university degrees. It should be expected that they have a significantly greater level of knowledge and understanding than the farmers they serve." Paul's

---

Numerous factors have been found to impact the effectiveness of the mixing process. These include:

- **Order of ingredient addition.** Small inclusion ingredients should not be added first. Portions may end up in pockets (e.g., at the ends of the mixer or in pockets around the discharge gate) where significant portions will not get mixed. Addition of liquids should be last or near to last.
- **Premixing/minimum inclusion rates.** Some ingredients are required in very small quantities. Direct addition to the mixer can result in poor distribution throughout these ingredients. Ingredients added to a single auger vertical mixer at less than 2% of the batch size should first be premixed with a portion of one of the major ingredients. When added to a horizontal mixer this can be as low as 0.5–1%.
- **Formulation.** Selection of ingredients and inclusion rates can affect mixing effectiveness. For example, ingredients differing in particle size and density will tend to segregate in a dry mix. Addition of small amounts of liquid like molasses can help keep small particles in suspension with particles of larger size.

Figure 5.4   Added tips to the manual section on "mixing." The section now contains more practical information.

opinion over the years had always been that teaching material should challenge even the more capable students in a group. "In doing so," Paul said, "all would learn more than would be the case otherwise, including those less able."

In the end, the three decided that to accommodate all audiences they would provide both thorough training materials and condensed versions with just the most important points. In the training manuals, this condensed version took the form of one- to two-page bullet point summaries at the beginning of each chapter. Trainees who were more capable could refer to the in-depth material contained in each chapter, while those who had difficulty with such details could focus on the bullet point summaries. An extensive glossary was also included to define and explain terms that may be more advanced or unfamiliar for some trainees. Additionally, manuals were translated into local languages to make the information more accessible. These additions created more work for Paul, Tesfaye, and Petros, but they also helped to ensure that all trainees would leave the training knowing more than they did before they came.

A similar approach was used in teaching the ag agents about formulation techniques. These techniques are used to determine which feeds, among available ones, are most economically priced based on nutrient content and how to combine them to efficiently meet an animal's nutrient requirements. The ag agents would be counted on to support farmers and co-op staff in routine decisions like what feeds to buy, how to combine them, and how they should be fed.

The tools taught to make these decisions ranged from simple to complex, so regardless of their capabilities, every participant would gain something useful from the training. With simple techniques, they introduced trainees to the "Pearson square." This is a mathematical technique that can be used to mix the correct proportions of two feeds so as to have the desired level of a single nutrient. The procedure is short (three steps), can be done with a pencil and piece of paper, and involves only rudimentary math (addition, subtraction, multiplication, and division). Figure 5.5 illustrates how Tesfaye and Petros discussed the Pearson square in the feed and formulation manual.

The advantage of this technique is that it is simple and quick and can be done by the trainers, agents, and farmers in the fields. Its main weakness is that it is only practical for combining two or three feeds and balancing for one or two nutrients.

To determine the optimum combination of many feeds and to meet the requirements for many nutrients at minimum cost, they introduced the trainees to complex techniques such as the least-cost formulation, which uses linear optimization modeling. Although the math is daunting, computers make the use of such applications relatively easy, fast, and routine. Ag agents usually have computers so they can provide the more complex data to individual farmers who would not have access to computers in the fields.

Although math is still math, the teaching of these complex techniques needed to fit the ag agents' available computer hardware and software, which is often old (e.g., 256k RAM, 80 MB hard drive, and Windows XP). Moreover, while many trainees had good command of the English language, some did not, which created a problem because commercial software packages designed specifically to carry out such computation tasks typically use English, Spanish, German, and French and cannot be changed. Moreover, most of these commercial software packages cost several hun-

Pearson Square. Feeds with their nutrient contents are listed to the left of the square. The desired nutrient level for the combination is placed in the center. Subtract diagonally across the square, always subtracting the smaller number from the larger. Total the differences and divide each difference by the total to obtain the percentage for each feed in the combination.

Example: Determine the combination of noug cake (31.7% protein) and tef straw (4.8% protein) necessary to create a ration that is 12.0% protein.

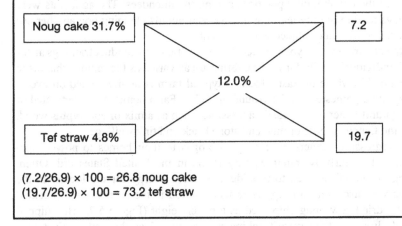

$(7.2/26.9) \times 100 = 26.8$ noug cake
$(19.7/26.9) \times 100 = 73.2$ tef straw

Figure 5.5   Explanation of Pearson square in training manual. This technique can be used to mix the correct proportions of two feeds.

dred to several thousand dollars per copy and are often more complicated than desirable for teaching purposes.

Considering these technical limitations, Paul, Tesfaye, and Petros decided that developing applications in Microsoft Excel was most practical. Although this meant Paul would have to commit many hours to the creation of Excel spreadsheets, they felt that the result would be more effective because the teaching could be targeted and the instruction could be translated into regional languages. An older version of Excel was used to create the applications so as to decrease the risk of compatibility problems with older computer systems used in the project area.

By December of 2009, Petros and Tesfaye completed drafting the training materials. Paul reviewed and edited the drafts, which was time consuming and even "painful" at times, but the review and editing gave Paul a better handle on what subjects Petros and Tesfaye were well versed in and what each was less comfortable or even unfamiliar with. The training materials were tested. Feed mix testing was done at a local vocational technical school that builds the mixers for the project. Petros and Tesfaye took their instruction sets and practiced using the ingredients and the machinery that would eventually be used at co-op unions and agriculture extension sites. They checked for order of operations, clarity of directions, and quality of the end product.

For the training sessions with the ag agents and co-op leaders, Paul, Tesfaye, and Petros had the benefit of using a computer laboratory classroom with 15 to 20 machines at a local college (Figure 5.6).

During the sessions, Tesfaye and Petros acted as the lead teachers and translators. Paul provided support by addressing the student questions Petros and Tesfaye were not confident responding to and clarifying needed information as necessary. Computer skills varied widely among the attendees. Some were proficient with using Microsoft Office and Excel, while others had limited abilities. As the sessions took shape, so did the collaborative practices among the attendees. The ag agents were actively engaged in not only learning the material but also helping one another use the software and translate subject matter content.

From the classroom, everyone proceeded to the "field" to conduct trainings at co-op sites or individual family farms. The farms are as varied as the regions that make up the geographic whole of East Africa. A typical farm is family owned on two to five acres with a homestead and surrounding fields. Farm homes are constructed of whatever natural materials are easily accessible, such as a mix of eucalyptus wood, tin, mud, and thatched roofs or mud and stone brick and tin dwellings.

Some progressive farmers have adapted biogas to their homes to power electricity, much like early twentieth-century farms in the United States did. Often, dried dung, a biofuel, is used to provide alternative energy for the locals. In the field, dung was also used to keep materials anchored in the breeze. Paul chuckled as he saw a dried cow dung chip used as a paperweight (Figure 5.7). He quipped that, hopefully, what they were teaching was as useful as the end product of a cow's intestine.

Figure 5.6 Computer laboratory classroom. This is the setting for the ag agent and co-op leader training sessions.

Figure 5.7   Dried cow dung chip. This chip was used as a handy paperweight.

TRAINING THE FARMERS

Once trained, ag agents would tailor the knowledge and skills they had acquired to the circumstances of the farmers in their own communities and communicate these practices to the farmers. Their materials had to be customized to meet the varying regional conditions and microclimates, and feed rations had to be adapted to meet local needs. For example, Rhodes grass is more drought tolerant and a better fit than Guinea grass for the relatively dry Rift Valley, but either would work in many of the East African highland areas where rainfall is greater and more dependable. When making molasses lick blocks, noug cake is more readily available in some areas than, for example, cottonseed cake, and appropriate adjustments need to be made to the formula. Plastic fertilizer bags work well for storing silage but are not available everywhere. Earthen pits (Figure 5.8) are then more practical in those locations.

Communicating these practices to the farmers was just as challenging. In East Africa, Christian, Muslim, and tribal customs and dialects vary by region. There are over 80 languages and dialects spoken in the Horn of Africa, too many to fully describe here. Most farmers speak only one of the four dominant Afro-Asiatic languages that are used in the regions. Consequently, before the ag agents could even begin, their presentations had to be translated into one or more of these local languages. Supporting materials, such as pamphlets and flip charts, had to be translated as well. The stark contrast between standard business English and typography on the one hand and Arabic calligraphy and Semitic language on the other made the cultural and lingual gaps apparent. Although adept at learning languages, Paul was aware that learning more than polite pleasantries was not the most useful support he could offer at this stage of the project. As he put it, "There's too much room for loss in translation." Simply putting the wrong guttural accent will change the tense or meaning of a phrase or word. With their multilingual skills, Petros, Tesfaye, and the

Figure 5.8 Earthen pits. In the absence of plastic fertilizer bags, earthen pits are used for storing silage.

ag agents or co-op leaders served as local trainers in order to help bridge the gaps. They customized the materials and presentations and trained the farmers.

The ag agents usually held trainings in local agricultural co-op facilities such as the newly constructed meeting hall in Figure 5.9 or at local café meeting rooms. Petros or Tesfaye would attend the sessions and take the lead as regional experts to support the ag agents. Paul would attend by invitation.

Figure 5.9 Farmer training session. Ag agents train family farmers in local agricultural co-op facilities.

The smallholder farmers would travel to town on foot, via donkey cart, and a rare few via bus or motor vehicles. The trainings would often coincide with market days to make the trip more convenient for the farmers. The number of participants per session was typically between 15 and 30. A meeting room would be filled with men and women, usually sitting as couples or in rows of separate gender groups. A few young women would be in modern western jeans and t-shirts while most of the older women would wear skirts, dresses, and scarves. The men in the sessions were almost all in work pants, shirts, jackets, and ubiquitous ball caps and a few wore a Fez-like head cover. Polite listeners in meetings, the farmers revealed their interest and concerns through their body language. Poised forward to listen, they would write down key points on the provided note pads. In general, however, only a few would speak during question-and-answer periods.

The participation of women was encouraged since women, many of them widows and single heads of households, operated many of the poorest farms. Female ag agents are especially appreciated by women in the Muslim areas where gender divisions prevail (see Figure 5.10). Such female presence was appreciated in whatever form it came. When Paul's wife came to a few of the sessions and talked to the women about her own dairy farm experiences, the male trainers gleaned more details about the needs of the women farmers.

Depending on the topic, trainings also included "hands-on" sessions where participants could learn by doing. This was especially needed because participants were oftentimes hesitant to speak up in a group, making it difficult to know how much they were really understanding. For instance, trainers wanted farmers to know that if a feed product used as a model in a session was scarce or too expensive, alternative feedstuffs could possibly be used. To help farmers with hands-on training, at the last stage of the project, ag agents followed the formal training sessions with less formal

Figure 5.10   Informal farm visit. A female ag agent (right) visits a female farmer (left).

farm visits to talk with the farmers one-on-one and determine just how well they understood the training material and the difficulties they might have putting their newly acquired knowledge into practice.

FOLLOW-UP VISITS

As Paul states, "In the end, none of this effort is of value unless it effects the desired changes." As the last stage of the project, follow-up visits by project staff to farmers participating in the project were conducted on a routine basis to assess the impact of the training. One such visit was paid in early December of 2009 to the farm of Ms. and Mr. Yetemwork. Their farm is approximately three miles from the nearest town. It is one and a half acres in size. Mr. Yetemwork manages the crops (i.e., tef, peas, beans, and sometimes wheat). Ms. Yetemwork is responsible for the livestock and forages. At the time, the farm had seven milking cows (five in lactation and two dry/pregnant), two heifers, and one calf. Milk production was being marketed to a local dairy processing co-op through a collection site in the nearby town.

Ms. Yetemwork attended four project trainings: one on forage production (including material support in the form of seeds/seedlings), a second on urea treatment of crop residues to improve feed value, a third on making molasses lick blocks, and the fourth on dairy cattle nutrition and feeding management. Subsequent to these trainings, she allocated a quarter acre of her farm ground to forage production (elephant grass, sesbania, and fodder beets were being raised). Most of the forage was fed to the cattle and a portion was being used for seed production. Previously this land was used for producing food crops.

During this visit, Ms. Yetemwork gave a private tour of her farm to Paul, Petros, another project staff member, and some visiting government officials. Ms. Yetemwork speaks only the local dialect so the other staff member, who speaks both the local dialect and English, translated for her and the visitors. Paul had told her earlier that the visiting officials were the source of the aid contributions, so Ms. Yetemwork was nervous and hesitant at first, but she handled herself well as she answered the various questions about the new farming techniques and their impact on production. She became more involved and animated as the questions and answers proceeded.

Ms. Yetemwork showed the visitors her dairy barn (Figure 5.11) and made a point about how milk production had increased due to the knowledge she gained by attending the trainings. "With adjustments in the rations fed to my cows," Ms. Yetemwork said, "milk production has increased from 10 liters to 12 liters for each cow in each milking." Four extra liters of milk a day from each cow represents a significant increase in income for a farm of this size.

Like virtually all other project participants, Ms. Yetemwork was very courteous and enjoyed showing visitors her cows and forage plot. She was proud of the progress she had made and went out of her way to make sure Paul, Petros, and the others understood how grateful she was for the assistance. "Your success is our success," Paul replied simply.

Figure 5.11   Ms. Yetemwork's dairy barn. Ms. Yetemwork's barn has higher producing modern dairy cows.

TECHNICAL KNOWLEDGE BUILDING: BRIDGING THE "CHASMS"

For Paul, Petros, Tesfaye, and other project staff, it seemed evident that effective communication of technical knowledge between groups and individuals from starkly different cultural, linguistic, educational, economic, and technological backgrounds was, by necessity, a collaborative effort. This collaboration required identifying existing communities of practice, forming and developing communities where none existed before, aligning project goals with the needs of these communities, and leveraging their practice to facilitate communication and training.

It was not possible for the NGO staff to reach the target audience or to create the appropriate technical training documents on their own. They had to rely on ag agents, co-op leaders, and individuals out in the field to help with localized differences in language and culture. Each stage of the process contributed an additional adjustment to the message that got it a little closer to one that could be understood and put to use by the smallholder farmers who needed them.

ACKNOWLEDGMENT

We owe a great deal of gratitude to Paul and his professional staff for their contribution, patience, and willingness to be part of this work.

RECOMMENDED READINGS

1. The term "communities of practice" refers to the development of a community that places learning (and writing) as an activity that is social in nature and comes largely from the experiences of participating in daily life. This concept can be particularly valuable for technical communicators to realize knowledge transfer, to facilitate communication, and to enact changes. Following are additional readings on communities of practice:

   • E. Wenger, *Communities of Practice: Learning, Meaning, and Identity.* New York: Cambridge University Press, 1998.

   • D. Barton and K. Tusting, Eds., *Beyond Communities of Practice: Language, Power and Social Context.* New York: Cambridge University Press, 2005.

   • L. Fisher and L. Bennion, "Organizational implications of the future development of technical communication: Fostering communities of practice in the workplace," *Technical Communication,* vol. 52, no. 3, pp. 277–288, 2005.

   • S. H. Regli, "Whose ideas? The technical writer's expertise in inventio," in *Professional Writing and Rhetoric: Readings from the Field,* T. Peeples, Ed. New York: Longman, 2003, pp. 71–78.

2. Recent research in development studies that focus on Africa and other places in the "Third World/Developing World/the Global South" increasingly acknowledge a serious problem with many development activities originating from or carried out by western governments, NGOs, and corporations. The problem has been labeled neo-colonialist. It involves a tendency to disregard local and indigenous knowledges, values, and preferences, and it presumes that western knowledge is the only true knowledge and western practices are the answer to every Third World problem. Alternative approaches are increasingly being advocated, which emphasize the inclusion of local and indigenous knowledges. Following are some sources that discuss these issues:

   • C. McFarlane, "Transnational development networks: Bringing development and postcolonial approaches into dialogue," *The Geographical Journal,* vol. 172, no. 1, pp. 35–49, 2006.

   • C. Sylvester, "Development studies and postcolonial studies: Disparate tales of the 'Third World'," *Third World Quarterly,* vol. 20, no. 4, pp. 703–721, 1999.

   • J. Briggs and J. Sharp, "Indigenous knowledges and development: A postcolonial caution," *Third World Quarterly,* vol. 25, no. 4, pp. 661–676, 2004.

DISCUSSION QUESTIONS

1. In the course of this project, learning, collaboration, and document production happened in various ways: in formally managed classes (such as the sessions

with ag agents in computer classrooms), structured teams (such as the brainstorming sessions among Paul, Petros, and Tesfaye), and during more informal practices (such as testing the instructions for molasses blocks in Paul's carport). In addition, written technical manuals are also more formal than oral instructions that may happen in the field. Consider these and other instances of learning that happened in the story. Discuss their respective structures, applications, and values at transferring technical information across cultural, linguistic, educational, economic, and technological "chasms."

2. Do you see communities of practice formed during this East African development project? If so, how did they form and what was their significance?

3. In what ways did people in this development project work communally and dialectically to realize knowledge transfer? On the other hand, in what ways was the learning that occurred tied to individuals' identities (such as the languages they spoke), their communities (such as the co-ops or the ag agencies), or the social practices that make up those communities (such as the roles of men and women)?

4. Go back and carefully read the explanation of the Pearson square. How clear is this explanation to you? Do an online search for Pearson square and read at least three different online explanations. Are any of them better than this one? After you have understood how to use the Pearson square, revise this explanation to make it clear enough to address any confusion you initially had with the one in this story. Perform usability tests of the story version, the three online versions, and your own version with students who are not familiar with the Pearson square. Is there a difference between explaining how to use the Pearson square and explaining the concept? Why might the concept be important? Why might it not be important? Would it be important in some contexts and not others? Explain.

5. The chapter puts quite some emphasis on the project manager Paul. Some readers may find this emphasis unsatisfactory and contend that Paul, as an older, white, western male, may eclipse the roles played by the local technical experts and farmers. This is a legitimate concern. However, as development projects are under the negotiation of NGOs and various local government agencies, the responsibility to include local or indigenous knowledge cannot be limited to project managers like Paul. In the midst of broader structural and social contexts—the expectations of the local government agencies involved, the goals and mission of an NGO, the needs of local communities, and aspects such as global economic, technological, and environmental factors—how did or how could Paul and/or other people negotiate the inclusion or exclusion of local/indigenous knowledges?

CHAPTER 6

THAT'S WORTH WAITING FOR: GLDD'S SPANISH JSA BOOKLET

Christina Schulz

Christina Schulz is 12 years strong in a career centered around collaborative learning environments, 10 of which she has spent at the Hile Group in Normal, Illinois. Schulz's consulting skills draw upon her previous life as a high-school English teacher in order to get groups thinking and learning in the moment. She brings lively, good-humored facilitation to organizational culture-change projects. And, with those skills, she supports participants' recognition of common ground and reliance upon consensus, which result in breakthrough solutions to organizational processes, including safety. In addition, Schulz promotes organizational learning and change through projects that require customers to work through perceived cultural barriers. For such projects, she helps customers engage in team-based writing and peer-led implementation, processes similar to the ones described in this chapter. In her down time, she learns, of course; current opportunities for this brand of fun include mentoring, parenting her children (Gwenyth and Collin), traveling, and getting into nature.

CHAPTER SYNOPSIS

In many industries, the way work processes are carried out can have life-or-death consequences. Christina is a consultant/writer for a performance consulting firm that specializes in workplace safety practices. But she was faced with a new challenge when she was asked to develop a job safety analysis booklet for hydraulic equipment operations for a dredging company in both English and Spanish versions. She had no prior experience in working with translators and learned that technical translators for this kind of documentation were hard to find. It was an especially difficult problem because the goal was not to develop the booklet in English and then to do a Spanish version but to release both versions at the same time. It emphasizes the importance of earning buy-in from the users through the booklet development process itself, not afterward. The unexpected solution to the challenge required establishing connections with Spanish-speaking members of the dredging community for translation, usability testing, and editing. Christina's story illustrates a philosophy that might be more widely adopted in intercultural and international technical communication.

———·•·———

Think of water. Land may be within sight or it may be many miles away (your choice). Now, also picture equipment with many moving parts; I won't force the technical names of equipment on you, but do think big. Dug-up silt, sand, or stone (again, your choice) from the water moves up and down, up and over, from here to waaay over there, through pipes, and in the mouths of forklifts and end loaders. Cranes move equipment. Rigging is tightened, then loosened. Belts whirl loudly—deafeningly—at high speeds. Diesel engines hum. And all the while, people come, go, bend, lift, straighten, pull, carry, drive, position, and adjust. Across the United States and worldwide, the interconnection of water, earth, machinery, and people pulses 24 hours a day, seven days a week. Dredgers pull from fresh and salt water alike pieces of the earth itself that get in the way of waterway transportation. In some cases, their work is to restore that silt, sand, or stone to where it is needed or wanted, and then dredgers become restorers instead of reapers. In either case, the interconnection between natural elements and human technology is ever dynamic, always experienced in the context of motion and movement.

There is a tremendous amount of organizational documentation to support this work, which makes for an intense combination of regulatory, manufacturer-driven, company-specific, and stakeholder-influenced standards, protocols, rules, procedures, recommendations, checklists, forms, and training tools. And that is just with a focus on English-language materials and resources. Imagine the intensity in going multilingual.

Great Lakes Dredge and Dock (GLDD as insiders call it) performs dredging services for the U.S. Army Corps of Engineers throughout the country and in Puerto Rico, too (see Figure 6.1). It also serves a number of private customers domestically and internationally, with about a quarter of its work in the Middle East. With over 300 employees and a hundred years' experience in the business, GLDD has access to many kinds of resources to operate effectively. The size and stability of the organization can also pose challenges to people and processes that seek to introduce change—any effort to introduce something "new" has to reach pretty far and wide to be completely implemented. But riding out those challenges can be worth it.

Think of the contrast here—I'm employed by Hile Group, a pretty small organization, what our local Chamber of Commerce calls a "microbusiness," and yet I'm attached as a safety/performance consultant and technical communicator to the

Figure 6.1 GLDD dredge. This is what a GLDD hopper dredge looks like in action. Image used with permission from GLDD.

pretty sizable company of GLDD. Any project I or my Hile Group colleagues work on has a ripple effect that may touch hundreds of people directly or indirectly. In one customer interface—through even just one project—I can reach twice as many people as I likely would have reached in an entire school year had I remained a teacher, which I was until 10 years ago. Take, for example, the dredging crew in Figure 6.2. I have not worked directly with them, but with them I have undoubtedly collaborated.

Figure 6.2 Dredging crew. A Bahrain-based work crew poses for a picture in safety gear. Image used with permission from GLDD.

JOB SAFETY ANALYSIS (JSA)

GLDD and I—well, Hile Group, that is—started to work together in 2007 to reinforce the company's commitment to safety, specifically on culture-improving efforts to go without injuries and incidents. By late 2009, David Simonelli, President of Dredging Operations, and the senior leadership team (SLT) had already worked with my colleague Julie Hile on a policy requiring JSAs for all routine and nonroutine tasks. The impact of that policy was that, for the first time, the company had decided its commitment to safe communication was strong enough to make pretask planning mandatory. Prior to 2009, pretask planning was deemed important but did not carry the weight of a requirement.

The process for completing a JSA includes identifying the steps needed to complete a task, the hazards associated with that work, and the countermeasures that have to be implemented to minimize the potential for personal or other harm from happening in the performance of an individual's or crew's duties (see Figure 6.3). This is also known in safety as job hazard analysis (JHA). The decisions to both formalize the process for completing JSAs and deem the process as a required safety skill set for employees generated the need to update the existing, first-edition safety job tools. I managed the project that resulted from these decisions and helped launch a series of second-edition JSA booklets. Here, let me show you how details related to a job task get analyzed to manage the potential threats to safety (see Figure 6.4).

JSA booklets such as this one carry the weight of a high priority for GLDD, rendering them a pretty significant instrument of the company's safety culture. Had you not known this, you would not have known their impact merely by looking at them: Thumb-printed or otherwise soiled pages with sporadic handwritten mark-ups in books only 3.75 inches wide by 6.25 inches long. There needs to be a corresponding booklet for each of the eight aspects of GLDD's operations: clamshell, drillboat, engine room, hopper, hydraulic, site engineering, tugboat, and yard.

In the Spring of 2010, I had worked for about four or five months under the

---

What is a Job Safety Analysis?

The Great Lakes Dredge & Dock Job Safety Analysis (JSA) is a process by which you and your coworkers plan how to perform a job task, identify the hazards associated with the task, and take the necessary precautions to complete the task safely, each and every time. Consider the JSA your "Game Plan" for working safely.

Great Lakes Dredge & Dock JSA Standards

1. Great Lakes requires that a JSA be conducted for all routine and nonroutine tasks.

Figure 6.3 Excerpt from GLDD JSA Requirements. This excerpt explains what JSA is about. Used with permission from GLDD.

Anchor Barge Moving

Required PPE

Recommended Tools: Hand Tools

Safe Job Steps

1. Take the time to stretch, either alone or with crew members, before completing the following job steps.
2. Confirm PPE needed and inspect before use.
3. Make tow with the anchor barge.
4. Move the anchor barge.
5. Secure the anchor barge.
6. Maintain housekeeping.
7. Stop work if you and/or the crew deviate from the JSA plan.

Hazard Analysis for Anchor Barge Moving

Recognized Hazards	Risk-Reducing Actions
Hands and fingers caught in rope	☐ Watch placement of hands and fingers. ☐ Maintain alertness.
Meeting traffic Grounding	☐ Talk to traffic by using VHF radio. ☐ Watch depth finder.
Pinch point	☐ Watch placement of hands while working. ☐ Ensure hands and fingers are out of pinch points.
Weather conditions	☐ Observe weather conditions. ☐ Stop work in lightning or severe weather conditions.
Slips, trips, and falls	☐ Clear all unnecessary items before working to keep decks clear. Stow equipment properly. ☐ Maintain housekeeping during and after performing work.

Figure 6.4   JSA from GLDD Hydraulic JSA Booklet, 2nd edition. Anchor Barge Moving is a representative JSA. Used with permission from GLDD.

joint direction of three senior managers at GLDD: David Simonelli, President of Dredging Operations; Glenn Thomas, Corporate Safety Manager; and Steve O'Hara, the VP of Clamshell Division and Corporate Incident- and Injury-Free (IIF) Manager. This latter role for Steve makes him responsible for safety-improvement efforts across the company, not just in his home division. We had successfully developed first-draft JSA booklets for six of the eight divisions (except for drillboat and engine room), which we referred to as phase I. Phase I culminated in a stack of completed Suggestion Forms listing recommendations for improving the drafts, written by employees and managers who had tested the books in the field during March and April. Phase II, then, included processing and adopting suggestions, developing the drillboat and engine room booklets, and moving toward final production of the booklets. The second-edition booklets I was to help produce replaced ones already in use, so we had some healthy pressure to make the forthcoming books more user-friendly to encourage a switch from the familiar old books to the new ones.

But then a twist in that early part of Spring—Steve indicated to me that we were going to seriously pursue a Spanish version of the hydraulic booklet, the division in which the highest number of English-as-Second Language (ESL) employees worked. There was no particular "Ta-Da!" moment to distinguish the kickoff to the Spanish booklet—Steve's personality is much more down to earth than that. It was more like, "Well, let's go ahead and translate the hydraulic book into Spanish," and the decision was made.

SEARCH FOR SPANISH TRANSLATOR

Receiving a request for translation was not news to me. The transportation industry, including rail, tow boating, and dredging, has for years been employing diverse crew members both in the United States and overseas. And I knew that GLDD has tried to offer multilingual employees information in their native tongues. Decisions about which language to use were based on the U.S. or overseas employees' requests. When I first worked with the hydraulic division in 2007, we launched a leadership development effort with the multilingual crew from the vessel *Alaska*, which included translating some meeting information into Spanish.

But it was the fact that the project had morphed into a *simultaneous* development and translation effort that struck me. I knew that most ESL employees at GLDD understand written and oral English well enough to have contributed to the English-version JSAs in use across the company, but they would also appreciate receiving the information in their native tongue. "We are going after English and Spanish together, at the same time," I thought to myself. The Spanish JSA booklet for the hydraulic division was both a safety job tool and a symbol of the company's recognition of ESL employees and the shared value and responsibility they possess for ensuring safety alongside their English-as-only language colleagues. Although supportive, I also felt self-consciousness nagging at me in the form of a repetitive question, "How will I help pull that off?"

You see, I can't speak or read Spanish. The French minor I earned from Illinois State University in 1999 has been collecting dust since then—as have the foreign-language realms of my brains since I settled into the postcollege world of working full time. But still, my earlier life has made me aware of the demographic needs within the U.S. workforce and abroad—not to mention the challenge of trying to become bilingual and how tough it really is to make an adult brain take on the ins and outs of a second language. I wanted to do a *dang* good job at this translation effort, suffice it to say. But I didn't know just yet where "Christina the tech communicator" fit into the effort. "Whoa, Chris," I remember cautioning myself, "this is a process, not a product. Don't forget that. The Spanish book will be one of a series, and it's the overall series that represents a new way of thinking and talking about safety at GLDD."

But I don't think the process manager in me had much sway over the budding monolingual Lone Ranger, and in hindsight, I see now that this is where I made mistake #1: I wanted to be part of a successful translation effort in a "look at me and my tech communicator prowess" kind of way. And, with this project developing text 100+ pages in length, I wanted to showcase this translation effort as being 10 times more powerful than what we did in 2007 for the bilingual *Alaska* crew members. I wanted a Ta-Da! moment associated with the JSA booklet. True, it was not a criminal offense. However, wearing the responsibility of translating the text like a kind of mantle across my shoulders set me up to be less likely to look to others for their ideas about how to best proceed or even hearken back to the well-developed process we have used in phase I to produce the other six books.

So, with the somewhat humbling acknowledgment that I wasn't going to be the translator, I was determined to be a critical stakeholder in completing the translation of something longer, more detailed, more critical to day-to-day safety for the Span-ish-speaking hydraulic division employees than what we did before for the bilingual *Alaska* crew members. Upon hanging up the phone with Steve after our brief project update call, in which he nonchalantly cued up the translation effort, I turned to a fresh page in my green-covered, college-ruled spiral notebook reserved for GLDD projects for the many questions I had about how to proceed with the request:

1. What is the starting point for finding a translator? The *Alaska* translation pro-ject was done with the help of a vendor to Hile Group, but what if the vendor is unable to do the work this time? (I have some doubts about asking a vendor who moonlighted for us previously to take on something as substantial as a 100+ page safety document.)
2. What will it cost to hire a translator for such a dense job?

I paused here, head tilted, gazing down at my project notebook, teeth idly nib-bling pen, and considered the text. Each JSA itself was a three-part item—job steps, hazards, and countermeasures, multiplied by the dozen or so topics that made up the core material of the book. On top of that, there was the JSA policy to translate, graphical safety icons to explain, write-in pages to support real-time development of new JSAs, and suggestion forms so that Spanish-speaking employees could recom-mend revisions to the booklet for a future edition. Not just lots of pages, but differ-ent kinds of information to manage in another language, too. So ...

3. How can I make the translator comfortable with dredging content?
4. How much time will a translator need to get his/her arms around all this work? And, how will that impact the timeframe for the overall project?

A little more pen-chewing: Hadn't Steve, who represented not just David and Glenn but also the overall team of subject matter experts (SMEs), indicated the equal importance of ESL employees in making the request for a Spanish booklet? That im-portance could be best demonstrated by a simultaneous launch of the English and Spanish books, no? And, if so, what would that mean for the finalized English books, including the two newcomers, if they were ready before the Spanish one was?

5. Won't it mean more to the crews if we are able to time the launch of the Span-ish JSA booklet to coincide with the launch of all the English language books?

Instead of sharing these questions with the very busy senior leaders who hired me to take on this project management task, I dove into independently exploring local translation resources, with zero luck. Our moonlighting vendor was tied up and un-available. Our local not-for-profit translation center housed in a community center actually ignored my phone calls, faxes, and e-mails after an initial phone call in which I was encouraged to send representative materials for review and considera-tion. (I never did learn the root of that sudden drop in communication ... not that I'm still grumpy about it or anything.) Then, I contacted Illinois State University's Foreign Language Department and, after another promising intro phone call, was gently turned down in my request for help.

Meanwhile, April slid into May, and May threatened to turn into June with no sign of a translator to partner with. And, to clarify to those of you outside of central Illinois, I work in a community with over 100,000 residents drawing from many backgrounds, levels of education, and professions. In addition to Illinois State University, Illinois Wesleyan University is here, State Farm is headquartered here, as are Country Financial and Mitsubishi's North American Manufacturing, and we have two large hospitals and the interconnecting medical practices that go along with them. Many graduates of Illinois State University stay in the community, and there are also quite a few retired people who stay here and continue to serve as volunteers, part-time workers, and so on. For all intents and purposes, there should not be a talent void in terms of finding a translator willing to work with safety materials like what we had been developing with GLDD. So, every promising start that ended abruptly downright confounded me. I kept thinking, "How is this going nowhere? What is wrong with our community?" In venting like this to my boss Julie, she recommended that I ask the GLDD folks about translating resources they have access to. The perfect opportunity arose during a June meeting with Steve O'Hara and several other critical SMEs on the book development project.

TAPPING INTO GLDD'S TRANSLATION RESOURCES

In early June, in Oak Brook, I facilitated the meeting around a long, rectangular table around which familiar faces sat—Steve, as mentioned, and other critical SMEs for this book development project. On the table itself were many marked-up copies of the field review English JSA drafts we had developed earlier in the year and a phone, which connected our in-person group to far-away SMEs. We roughed out plans for making final drafts out of the already-vetted booklets, prepared to get engine room and drillboat booklets up and running by July, and then turned to the question of the status of the Spanish JSA booklet. "Time to use Julie's coaching," I said to myself.

I took a big gulp and said regretfully, "So far, I haven't had much luck finding a translator. Julie reminded me that GLDD may have translation resources, particularly at your overseas locations, that may be called upon. Can you let me know if you have some ideas about that? We'll want to get this underway as soon as possible to keep the timeframe for launching this booklet in line with the English books."

And what I thought would happen didn't happen. No one looked at me critically or even hinted that I owed them a neatly packaged resource with the entire translation completed (yesterday). Apparently, I didn't sound like an ignoramus or incapable project manager and technical communicator. Instead, they took it as an important challenge for the team to chew on, to help resolve, so that the project could go forward, with me very much positioned to continue stewarding the work at the detail level once a translator could be located.

The team sort of tilted its collective head and thought for a moment. "Okay, we'll look into that," said calm-and-cool Steve O'Hara.

"I'll look into that," said resource-ready Glenn Thomas. And the meeting moved on productively for the rest of the afternoon.

Sigh of relief ... short-lived sigh of relief. Back at the office, with June continuing to progress and no follow-up from the team, I put on the monolingual Lone Ranger suit again and did something I didn't think I would ever do: cold call to translation service providers I had found through ads online. They were not helpful and, in fact, reminded me of why we use the term "cold" to describe calls like that in the first place. Neither the sole practitioner based near Chicago or the large outfit based in Chicago was able to take on such safety-specific technical work. For the sole practitioner, the content was too unfamiliar, resulting in a lack of interest in taking on the project, maybe spurred on by a lack of confidence (assumption there, admittedly). The larger organization failed to follow through, which, as another assumption, I interpreted as 50% likelihood that my request didn't match skill sets and 50% likelihood my request was buried under many other similar requests. By the time I completed these last two translator-hunting expeditions, I really caught on to trusting the team in its offer to help find the translator. By continuing on my solo search for "the one," I was potentially, fatefully, jeopardizing the *process* of how the Spanish JSA booklet would come into existence. Unlike the six, soon to be eight, English language books, the Spanish hydraulic text could lack the essential "of the people" soul, the soul of the people who do the work. Handing off to a third-party stranger meant accepting a third party's interpretation of the work. It would be someone's assignment to translate, count the words, generate the per-word invoice, and so on, with no connection to the spirit of either safety or the employees at GLDD. That did not feel okay.

I was in a quandary. As a non-Spanish-speaking person, how could I *make* the translated text live up to all that I knew the English language books represented?

Then, just as we were about to reach July, I received an unexpected phone call from Glenn Thomas. He was one of the drivers of the overall JSA booklet development, and I thought it was going to be similar to the early-June meeting, a call in which we compared notes on our progress toward completing the English-version booklets, confirmed that hard-copy printer estimates were forthcoming to anticipate final-book costs, that kind of thing. Well, I was wrong.

After our usual warm hellos and how-have-you-beens, Glenn jumped right in with news. "I think I have found the solution to the request you made about finding a translator for the JSA booklet. I wanted to run it by you."

Hooray! Although I was professional enough not to say that aloud. "Sure, that's great! I've really struggled to find a Spanish translation service that will take on both dredging content and a safety focus. JSAs are more technical than the average translator can handle. I had no idea it could be so hard to find a translator, either here or up in Chicago. It was really bugging me."

"I have been asking around within Great Lakes, and I learned that a couple of our dredgers who live in Texas are married to teachers. So, we could work with one of their wives to translate the hydraulic JSA booklet. What do you think?"

"Glenn, I love that idea. How do we get in touch?"

"I'm glad you think it will work, Christina. You can send information to me, since I still need to work out the arrangements within Great Lakes."

Upon hanging up the phone, I exhaled a stress-relieving sigh of relief that I could feel clear down to my toes. I sensed that Glenn's instincts as a long-time member of the GLDD safety culture were exactly right. A translator from the GLDD family, literally, would be the best of all possible worlds. She would literally translate her care for her husband's safety into a hard-copy resource that would support and protect not just him but the whole crew. And, with employees knowing the *who* behind the text, the effort could be more easily supported and endorsed. It wouldn't be the *company's* Spanish safety job tool. It would be *theirs*.

EUGENIA'S TRANSLATION

In late June, the English hydraulic booklet was ready to go to the translator, Eugenia Salinas, wife of GLDD's Lester Salinas. All Phase I suggestion form edits had been made. I e-mailed a brief memo indicating the readiness of the draft and appreciation for Eugenia's willingness to take on this task. I attached the finalized English JSA booklet for the hydraulic division in Microsoft Word, already formatted in the 3.75 × 6.25 size. Eugenia needed no warm-up but began steadily translating line by line all text from her home location in Texas. When and how often she pulled on Lester and his colleagues for context and explanations of dredging I do not know. The home-grown Spanish JSA booklet was no longer my plant to tend—just a seed I passed along to the right green thumb. Eugenia was light on communication about the translation process and heavy on the translation itself. Since Eugenia had what she needed to work with—both in materials and Spanish comprehension—I was left to wait, which my internal process manager appreciated while my monolingual Lone Ranger resisted. Before the Lone Ranger suit had another chance to come out of the closet, however, I received this forwarded e-mail from Glenn, with whom Eugenia primarily contacted, since they already knew one another through Lester:

> Mr. Thomas, attached are the files that you sent me translated, they have the same name I just added the word Spanish, so you know the difference, also I kept the same format as you asked, please review them, and if there is something you need to change just let me know.
>
> Thank you,
>
> Eugenia

And there it was, maybe a week after making the request or so, a Spanish text for hydraulic division employees. "This is *soooo* not about me," I muttered, half in amazement, half in rebuke.

In July, preprint proofreading of the hydraulic division English JSA booklet brought a couple of edits to light that meant making likewise edits in the Spanish booklet. And these prompted me to communicate directly with Eugenia:

Me:

Hi, again!

One quick question. What is the Spanish translation for "Sling Capacities"? It references a chart that tells you how much a sling of a crane can bear weight-wise.

Thanks! Getting closer to being done!

Have a safe weekend.

Christina

Eugenia:

Hi, the translation for this phrase is, "virutas, cascarillas o particulas."

Have a nice weekend.

Eugenia

As quickly as they did before, answers seemed to magically appear in my Outlook inbox without acknowledgment of how the work was being generated. It just was. Eugenia was working within an employee-centered (or spouse-of-employee-centered) safety effort without talking about it as a process, the way I was as the technical communicator on the job.

That didn't mean, though, that she wasn't using a reliable process for completing the translation. In one instance, I got a hint that she was. There were certain changes being made in other divisions' booklets, changes that crossed into two sections of the hydraulic book: "Crew Boat Personnel Transfer" and "Crane Rigging/Lifting." So I was making updates to the English version of the hydraulic book to make its safety standards identical to the other books. As a result, the Spanish translation needed to be changed, too. But having gone through months of niggling project details, I had a lapse in memory and forgot one niggle: Did I send the updated English book to Eugenia? Confined to my monolingualism, I had taken it for granted that a spot checking of work would take only seconds. But now, to ensure correctness in a foreign language, I had to learn to be patient, to rely on others, and to invest more time to do request making and waiting. I did my best to be patient, secretly hoping for Eugenia's proven timeliness. Lucky for me, within a day, I had that poof! in my e-mail inbox that had taken me by happy surprise previously.

Eugenia:

You wrote in your e-mail that for Crew Boat Personnel Transfer you had a new #5, I translated and added that one, and on the request to take out rows from the topic called "Crane Rigging/Lifting" I took those out because they were still in there. Anyway, if you have any questions or comments, please let me know, have a great week!

In my mind's eye, I could picture Eugenia in a home office, desktop computer fired up, with the Spanish language JSA booklet in Word open before her, reading my English language e-mail, and comparing those requests for review to the Spanish language text before her. A bit of scrolling with the mouse, a little reliance on the delete button, and *done*. The document was ready to resend to me, with a brief e-mail indicating what changes were made. I imagined her with a similar head tilt as my own, but I was confident she didn't need the anxiety-relieving activity of pen chewing, at least not then.

I never did meet Eugenia or talk with her by phone. We exchanged e-mails and Word documents—politely, efficiently, effectively. Our strongest common denominators—friendliness packaged in specific, time-sensitive messages, commitment to sending GLDDers home safe every day, and shared sense of urgency for timely project completion, all equally important in a solid working relationship spanning hundreds of miles—generated a feeling of interdependence to the translation effort that I had not expected and, certainly, had not felt when operating from the monolingual Lone Ranger mindset.

Figure 6.5 shows the same pages in Spanish given earlier in English in Figure 6.4 so you can appreciate the translation.

THE SEARCH FOR TRANSLATION REVIEWER

Eugenia's timeliness, and maybe even her embedded process for ensuring quality, ensured that the Spanish book could go out at the same time as the English books to show that the needs of bilingual employees were anticipated at the same time as English-speaking employees. The project was winding down.

And this is where I made mistake #2, which came upon me, not like a ton of bricks, but more like that slow-heating pot in that proverbial frog metaphor: Although Eugenia had completed her work on time, we had not identified Spanish-speaking employees who could give the draft the same read through and user test that all other books had received. This was out of step with what we did during phase I; employees were allowed the opportunity to carry drafts of the booklets around the workplace to "test them out" for a while before considering the books complete. Once again, I had been operating too narrowly in my role as project manager, focusing on the translation task, forgetting to situate the translation into the bigger picture of the book development process. In addition to existing, the Spanish JSA booklet needed to have credibility within the Spanish-speaking workforce, and the mere knowledge that a within-GLDD-ranks translator did the translating would not be enough to make people rely on the booklet every day.

I again flipped open the project notebook and began listing the questions I had about finalizing the translated draft:

1. Who are the employees who will take on the review of this draft? Where are they (and in what time zone, in case I need to talk to them by phone)?
2. Will the employees be on duty or between hitches (term used for someone who works an on/off schedule—many dredgers work 30 days on, 30 days off,

Movimiento de la Barcaza Ancla

Equipo de Protección Personal Requerido (PPE)

Herramientas Recomendadas:
Herramientas Manuales

Pasos para un Trabajo Seguro

1. Tómese el tiempo para estirarse, ya sea solo o con otros miembros de la tripulación, antes de completar los siguientes pasos.

2. Confirme el Equipo de Protección Personal que se necesita e inspecciónelo antes de usarlo.

3. Remolque con la barcaza ancla.

4. Mueva la barcaza ancla.

5. Asegure la barcaza ancla.

6. Mantenga la limpieza.

7. Detenga el trabajo si usted y/u otro miembro de la tripulación se desvían del plan original del Análisis de Seguridad del Trabajo (JSA).

Análisis de Riesgos al Mover la Barcaza Ancla

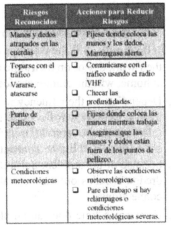

Riesgos Reconocidos	Acciones para Reducir Riesgos
Manos y dedos atrapados en las cuerdas	❏ Fíjese donde coloca las manos y los dedos. ❏ Manténgase alerta.
Toparse con el tráfico Vararse, atascarse	❏ Comunicarse con el tráfico usando el radio VHF. ❏ Checar las profundidades.
Punto de pellizco	❏ Fíjese dónde coloca las manos mientras trabaja. ❏ Asegúrese que las manos y dedos están fuera de los puntos de pellizco.
Condiciones meteorológicas	❏ Observe las condiciones meteorológicas. ❏ Pare el trabajo si hay relámpagos o condiciones meteorológicas severas.

Análisis de Riesgos al Mover la Barcaza Ancla (Continua)

Riesgos Reconocidos	Acciones para Reducir Riesgos
Resbalones, tropiezos y caídas.	❏ Quite todos los objetos innecesarios antes de trabajar para mantener la cubierta despejada. Guarde el equipo apropiadamente ❏ Mantenga la limpieza durante y después de realizar el trabajo.

Figure 6.5 JSA from GLDD Hydraulic Spanish JSA Booklet, 2nd edition. This is the Spanish translation of the Anchor Barge Moving JSA shown in Figure 6.4. Used with permission from GLDD.

135

for example, so they work seven days a week when working and then are home for about the same number of calendar days as they worked)?

3. If on duty, will the employees be able to review a heavy set of papers during their work day or on their time off (which needs to include rest, obviously)? If at home, will the employees feel like doing pen-and-paper work like this?

4. Depending on their fluency in English and their opportunities to talk during their work day, will I be able to understand them well enough to take down their requests for edits? (Dredging is *loud* work on/near water. Even the clearest of clear cell phone connections challenges the speaker and listener alike because of the ambient noise that accompanies the conversation, and when a dredge is miles out on the water, there isn't the clearest of clear cell phone connections.)

5. How much time does an employee need to digest all the pages of content? And, how might that impact the timeframe for the overall project?

6. What will we learn about the print readiness of the draft that Eugenia prepared? (I was unable to tell even if there were spelling errors in the book. Our abilities to format the Spanish book identically to the English one only reinforced my admitted sense of false security. It looked right, but was it?)

7. What will the consequences be if we launch the English books ahead of the Spanish one? What would the other departments say about waiting on launching their English language books until Spanish is ready to go?

8. What is the right amount of contact for an external consultant to have with customers? Where would I need to draw a line to limit my reminders (and possibly nagging) and just let GLDD managers' follow-through influence the resulting timeframe for completion? As an outsider, would I be able to accurately assess whether delays are due to legitimate operational reasons (i.e., particularly busy time for GLDD or change in priorities moving this project down a notch or two on the to-do list) so as to not appear insensitive in treating this project as *the* top priority because it is *my* top priority?

Again, I primarily kept these questions to myself, not wanting to overburden either my bosses on the project or the writing team I was collaborating with (yes, I, too, can appreciate the tinge of irony of a communicator withholding communication). Instead of sharing the above list, I compressed all my questions into an SOS request to Steve and SMEs from the hydraulic division to get some employee-level eyes on the text Eugenia had translated before it went to the printer. Again, I received a "We'll send some suggestions your way" response from Steve, and I generally trusted help would come—eventually. But, just as I did early on in handling the translation request, I felt overwhelmed by the feeling that I was somehow alone in my convictions about getting this draft user-tested, that the urgency in timeliness was not shared with the team, and that the ultimate outcome—successful launch of *all* books—was squarely on my shoulders to protect, even though I did not have a network with Spanish-speaking employees at the ready and was not positioned within GLDD in such a way that my physical presence would act as a reminder on follow-ups owed to the project.

At last, though, I made the call that finally created a breakthrough to this bottle-necked issue. After several rounds of e-mail tag with the Hydraulic Division Dredge and Division Safety, Health, and Environmental managers, I had a name: Cesar Silva.

Sigh of relief mixed in with a bit of "Duh!" blaring in my head. Yes! I had enjoyed his participation in the *Alaska* crew's leadership development process several years ago and remembered that, as deck captain, he helped ensure that our session was making sense to the Spanish-speaking deckhands we had in the class. It would be great to get his eye on this work.

But, as you likely have guessed, it was yet another short-lived sigh of relief. That quick burst of excitement as I envisioned project movement and working with someone I had known and respected transitioned to a protracted deflation, like a bicycle tire slowly losing air. Tracking down an individual dredger using cell phone and e-mail contact information *is hard*. Dredgers work rotating schedules, which can range from 6 to 12 hours on-duty to 6 to 12 hours off-duty. Hydraulic division employees work 12-on, 12-off, which affected my window for being in touch with Cesar, since I did not know whether his on–off shifts were day/night or night/day. That window was, potentially, pretty narrow, considering all that had to be coordinated: Cesar's ability to check messages during breaks, the online connectivity where his project was occurring, and timing a return message within my awake hours, either through a call at the office during the (landlubber) business day or at home using my cell phone number. Deck captains have time enough to honor required communication—logs, forms, reports, and, of course, whatever connection to the home front is possible. So, it wasn't clear to me where a request to do something above and beyond regular job duties would fall in terms of ability to follow up *according to my sense of timeframe*. I had no doubt Cesar could do a fabulous job; it was, at that point, about coordinating his eye on the text with my sense of time pressure. And, unfortunately, Cesar called my office twice when I wasn't at my phone. And although he left messages each time, he did not respond to any of the questions I had asked in my messages to him.

What I did learn in this communication loop with Cesar was that he was on duty and he was interested in helping out, but it wasn't clear whether he had printed out the info I had sent. So, it could be that my reviewer had nothing to review. Even if we did connect voice to voice, would there be feedback?

Eventually, something triggered a memory that I had a contact list from 2007—I had other people within GLDD who could get me through to Cesar. I could contact a contract manager who had both a shoreside office and daily contact with the dredge and all the people involved in the project. That person would be able to advise me on the best time of day to reach Cesar and the ground-mail address for getting the pages to Cesar in hard copy (taking one to-do off his shoulders by not having to make printouts of e-mail attachments). Although pressed for time, I took time to quickly grumble to myself, "Why hadn't I thought of that before?" I found the contact list, identified a contract manager with whom I had previously worked and liked working with, and wasted no time phoning him for help.

"Hi, Dave. Sorry to bother you, I know you are always on the go. Can you help me be in touch with Cesar? I understand he'd be the ideal SME for our Spanish JSA booklet. This is a time-sensitive request for Cesar's eye as a reviewer. I've tried to e-mail him the info and to coordinate via voice mail, but no luck."

"I know that he has tried to reach you," Dave replied, "but it's going to be tough to get access to him. He's working long days, and the dredge is offshore. Have you tried to call the dredge directly? Have you tried to e-mail Fidencio Gonzalez? He's a production engineer who also speaks Spanish. He has more access to the computer during the day."

"Dave, you're great. In five minutes, you've told me more about Cesar's availability and alternatives to reaching him than I have in the past two weeks of attempts to make contact. I'd be very glad for the number to the dredge, and I would welcome Fidencio's additional set of eyes on the Spanish documents."

CESAR SILVA'S FEEDBACK

With two new phone numbers, as well as the e-mail to another bilingual GLDD employee, I was now reenergized in getting this proofreading loop completed. We were definitely nearing the finish line, and everyone attached to the rollout of the project looked forward to launching all nine books together (eight English and one Spanish) to present one unified GLDD voice on the topic of routine JSAs—a voice employees would engage with and respond to. After a brief call with Fidencio in which I introduced myself and asked permission to sort of use him to get to Cesar, I quickly sent off an e-mail with the official request for review and corresponding attachments.

Me:

Hi, Fidencio!

As you know, we've been coordinating with Cesar to get his eye on the translation we developed to the English Hydraulic booklet. Dave Johanson recommended you as someone who may have helpful feedback for the Spanish JSA Booklet. Would you have the time and interest in taking a look and letting us know if anything is amiss terminology-wise? I can add your two-cents to what Cesar has offered, which would validate the translation we've done. FYI, our translator is actually the wife of a GLDD dredger, so we're lucky to have access to someone familiar with the industry and committed to reinforcing safety.

Some questions I have:

1. Do Spanish-speaking employees use Spanish or English terms for "Captain," "Leverman," etc? Do they work partly in English and partly in Spanish for referring to other crew members or anything? We should have the booklet written in a way that matches how most Spanish-speakers talk.

2. We've been putting topics in alphabetical order so that people can

find the topic they want easily. Hoping that will be good for Spanish-speaking employees, as well. But we want to make sure of that—our other option is to put topics in the Spanish booklet in the same order as the English one (although that won't necessarily be easy to work with).

Fidencio, same day:

Christina,

I will get with Cesar later on today as his shift ends around 6 pm, but they have been very busy with dredge repairs. I will be more than glad to help out in reviewing the booklet except that I just started working here and I'm learning the wording for the dredge industry. I will try to get with Cesar today to go through the booklet and add anything to make it more useful. I know for a fact that they have been working late and I'm not sure at what time he will be available. Having it ready by tomorrow may be a little hard, but we'll try our best.

This quick response provided great reassurance and prevented me from additional rounds of phone tag with the dredge on which Cesar was working—so many outgoing calls were made to coordinate operations, it could be fairly challenging to get an incoming call accomplished. I sent along another e-mail, hoping that Cesar would provide me with not just a word-by-word read of the safety information but also some sense of how ESL employees completed their own safety planning while remaining a part of what was occurring among English-speaking colleagues.

Fidencio, next day:

Christina,

On behalf of Cesar Silva, he has looked over the booklet and everything looks great except that some of the wording that is properly translated to Spanish may not be understandable in the field as some of the dredge crew may call it different. For example, on page 10 the word "Sanco" for spud is typically called "Pata." Cesar mentions that as long as the person in charge of the JSA understands what he is analyzing it is perfect. Other than that it looks great. Please feel free to contact Cesar if you have any questions.

Cesar called you on Friday September 24th, but wasn't able to reach you so he left you a message.

It was a relief to know that someone as well-respected as Cesar felt prepared to stand by the Spanish JSA booklet. Because he worked closely with the entire crew, he would undoubtedly anticipate how others would be able to work with the information and add that perspective to his own.

And yet, as you may have noticed, the word-level help Cesar provided had not entirely satisfied my curiosity about how Spanish-speaking employees would use the booklets alongside their English-speaking crewmates. Another SOS to Steve as project lead with a final question: Whose needs are most important here? The need for Spanish-speaking employees to find the topics according to alphabetical order, or the need for non-Spanish speaking supervisors to ensure Spanish and English employees alike were on, literally, the same page for safety by organizing the contents in identical order in which case it would be alphabetized according to English but not Spanish. "Spanish-speaking employees," was Steve's true-to-form simple, straightforward reply.

And, just like that, we were ready to go to print, English and Spanish together. Period.

Once I had stopped operating under "rules" I thought I had to adhere to as project manager of the JSA booklet effort—as well as let go of a sense of "noble mission" that kept encouraging an internal superhero to emerge where she was not needed—I realized all along I had the network I needed to be in touch with Spanish-speaking employees. With a more than three-year relationship with GLDD, I had met and worked with many great people, most of whom, like Eugenia, Cesar, and Fidencio, were only too happy to be called upon to help the company get better at something, particularly at safety. Somehow, I had compartmentalized myself as technical communicator for JSA booklets, and I forgot that no matter what role I was playing at present, I still had all kinds of people to draw upon—the extended team of managers, dredge operators, safety consultants, and several fully invested translators. And not a Lone Ranger among them!

RECOMMENDED READINGS

1. To learn more about the issues related to creating, translating, and localizing information for Spanish-speaking audiences and in the Latin American context, refer to the following sources:

   - N. St. Germaine-Madison, "Localizing medical information for U.S. Spanish-speakers," *Technical Communication,* vol. 56, no. 3, pp. 235–247, 2009.
   - N. St. Germaine-Madison, "Instructions, visuals, and the English-speaking bias in technical communication," *Technical Communication,* vol. 53, no. 2, pp. 184–194, 2006.
   - B. L. Thatcher, "Writing policies and procedures in a U.S./South American context," *Technical Communication Quarterly,* vol. 9, no. 4, pp. 365–399, 2000.
   - B. L. Thatcher, "Intercultural rhetoric, technology transfer, and writing in Mexican maquilas," *Technical Communication Quarterly,* vol. 15, no. 3, pp. 383–405, 2006.
   - E. Tebeaux, "Designing written business communication along the shifting

cultural continuum: The new face of Mexico," *Journal of Business and Technical Communication,* vol. 13, no. 1, pp. 49–85, 1999.

2. The following sources discuss, more broadly, how to prepare technical communication students and practitioners for translation and localization work:

   • R. A. Davis, "Nine things your translators wish you knew," *Intercom,* September/October, pp. 8–9, 2006.

   • B. Maylath, "Translating technical documents," podcast for the IEEE International Professional Communication Conference, July 13–16, 2008, Concordia University, Montréal, Québec, Canada.

   • B. Maylath, "Writing globally: Teaching the technical writing student to prepare documents for translation," *Journal of Business and Technical Communication,* vol. 11, no. 3, pp. 339–352, 1997.

   • B. Maylath and E. Thrush, "*Café, thé ou lait?* Training technical communicators to manage translation and localization," in *Managing Global Communication in Science and Technology,* P. Hager and H. J. Scheiber, Eds. New York: Wiley, 2000, pp. 233–254.

   • K. St.Amant, "Designing effective writing-for-translation intranet sites," *IEEE Transactions on Professional Communication,* vol. 46, no. 1, pp. 55–62, 2003.

   • K. Brown-Hoekstra (2011, Dec.), "Integrating localization and technical communication: 10 critical tasks," *tcworld.* Available: http://www.tcworld.info/tcworld/translation-and-localization/article/integrating-localization-and-technical-communication-10-critical-tasks/, accessed May 8, 2012.

DISCUSSION QUESTIONS

1. Christina's story illustrates some of the internal and external tensions that challenge a technical communicator working on a translation project. What are those tensions and what additional tensions can you foresee?

2. What do you think of the process through which the Spanish language JSA booklet was made possible? Are there things that you wish Christina and other characters in the story would have done differently? Why?

3. Which term do you think best applies to the communication challenges Christina was dealing with in this story: translation, localization, or internationalization (for more definitions of these terms, see Chapter 1)? Would you apply a different term at different points as the story develops? Explain your answers to these questions.

4. Despite its share of challenges, what value did this project have for Christina, for the other characters in the story, and for the various organizations and communities in the story?

5. Regarding the Spanish language JSA booklet, what do you think or what

questions do you have about its impact on employees and their safety? What would you do or whose help would you ask for to get information about the impact the translation is having?

6. As you think ahead to projects you will have in the future, which of your professional values may come into conflict with one another in an intercultural, multilingual environment? What might you need to do to bring them into better alignment and cooperation?

CHAPTER 7

MAKING EXCELLENCE A HABIT ACROSS INTERCULTURAL BARRIERS

Saurabh Kudesia

Saurabh Kudesia is currently working with Microsoft India Research and Development Pvt Ltd at Hyderabad. He has experience of over a decade working with globally distributed documentation teams and providing documentation services for a variety of domains. In his previous jobs, he managed 120+ documentation projects and implemented over a dozen medium/ large-size knowledge bases for IT, Information Technology Enable Service (ITES), Business Process Outsourcing (BPO), and product-based companies/ clients globally.

Prior to joining Microsoft, he was a member of the panel of expert evaluators revising ISO/IEC 25612 standards and made significant contributions to the Unicode Common Locale Data Repository (UCLDR) project. He cofounded and later served as editor-in-chief of KnowGenesis International Journal for Technical Communication (IJTC). He also provided his services as associate editor to Directives, a newsletter published by the Society of Technical Communication (STC)'s Management Special Interest Group (SIG). He has contributed multiple works in international research publications and conferences and has conducted several workshops globally in subjects such as content management, knowledge management, and documentation process improvements.

In addition to holding an engineering degree in electronics, Kudesia is a Certified Scrum Master (CSM) and an alumnus of Symbiosis Institute Pune and Indian Institute of Management (IIM) Bangalore. He can be contacted at saurabhkudesia@gmail.com.

CHAPTER SYNOPSIS

As globalized as today's business has become, companies often operate across national borders. In this story, which is told in the third person and based on the writer's personal experience, we watch the young Chinese telecommunication company XinZin growing quickly into a global enterprise. As overseas business boomed, so did the demand for quality technical communication in the English language. Sanjeev, a "foreign" manager with both technical communication and international experience, was handpicked to manage XinZin's documentation effort. Although XinZin was, on the surface, an internationalized company with global business and dispersed project teams, deep down, it still had many traits of a traditional Chinese company. Sanjeev realized this a little too late. Before he knew it, he was fighting a hard battle against language barriers and deep-seated documentation practices, working with a suspicious boss and bitter team members, and trying, in vain, to satisfy international customers. It was a chaotic and stressful experience for all involved until a cross-cultural training program provided a glimpse of hope.

DEDICATION

Dedicated to my two-year old loving son, Nachiket Kudesia, who left for the heavenly abode on 26 February, 2011.

The 20 faces inside the closed doors of the boardroom were helplessly staring at each other. XinZin's CEO was sitting quietly after dropping a bombshell a few minutes before. (All company and character names are pseudonyms.) Now the tersely worded contract cancellation letter from XinZin's prospective Japanese partners glared accusingly at them from the large projector screen.

RISING STAR

In the early 1990s, a group of Chinese-born, European-educated entrepreneurs set up XinZin, a Beijing-based company that sold telecommunication products to telephone operators and helped with the installation and operation of low-cost multimedia services. XinZin went public in the next four years, followed by a period of phenomenal growth that brought the company fast-rising earnings, a soaring stock price, and a new line of superior products. Within 10 years, XinZin was controlling 65 percent of the Chinese network gear market while continuing to expand in major global markets. Riding high on a series of acquisitions that enabled it to manufacture and sell equipment for high-speed Internet access and other value-added services, XinZin took a giant leap in its research abilities and established new development centers in Shanghai, Suzhou, New Jersey, and Bangalore. With $5 billion in revenues, a 6000+ strong workforce, and multiple sales offices across the globe, XinZin was close to becoming one of the leading equipment providers in the telecommunication industry.

Lurking under this seemingly ideal condition, however, were serious challenges: greater complexity and differentiation of international units; the need for better integration and alignment of people, process, and technology; and the problem of transferring knowledge and innovation across its global divisions. Contrary to its customers in mainland China, XinZin's international customers were far more demanding on quality issues. One of the biggest complaints of prospective buyers was that XinZin's products had virtually no supporting documents. At a time when XinZin's competitors were offering superior, comprehensive, and detailed documentation packages, XinZin was offering its customers either incomplete and inconsistent product manuals written in Mandarin, Cantonese, or a mix of the two or poorly translated pieces of technical documents in English that were difficult to comprehend. These supporting documents were unable to meet the customers' quality compliance requirements. The customers wasted a great deal of time coordinating with XinZin's support teams to resolve routine issues that could have been addressed in the product manuals. Most of the time the problem remained unresolved even after several follow-ups, leading to growing customer dissatisfaction.

When the prospective Japanese company pulled back from a lucrative offer to purchase XinZin's products due to lack of product documents and quality certificates, XinZin's management could no longer ignore its substandard documentation. Within a few weeks of losing the Japanese contract, XinZin CEO announced the formation of the Product Development Group (PDG) under the leadership of Sabrina, Vice President of the Product Engineering Group (PEG). An efficient administrator, Sabrina, who was in her early 50s, had been with XinZin for more than a decade and had an impeccable reputation in the Chinese industry for her charismatic leadership. Helping lead the PDG team was Xaoxing, General Manager of the Software Development and Testing Division.

Other managerial and team lead positions were filled. Jao was the senior manager responsible for the documentation of multimedia products, and Meng was the senior manager responsible for the documentation of common functions. Meng and Jao had in-depth knowledge of XinZin's products, but their exposure to documentation practices was limited and it was their first experience working as full-time documentation managers. To scale up the capabilities of Chinese documentation teams, Xaoxing hired Peter, from the U.S. and in his mid-40s, to be the senior manager of quality control. Peter was a versatile technical professional with many years of experience in the global telecommunication industry. Xaoxing also hired Tom, an Australian, to be a full-time editor. Lastly, Suzanne, a seasoned technical writer with extensive management skills, was made senior manager of international documentation at the R&D center in New Jersey. Most of the team members were based in Beijing, and additional technical writers were hired in Bangalore, Suzhou, Shanghai, and Dubai to support the activities targeted for Asian and Middle East markets.

PDG thus had all the characteristics of a diversified and distributed global organization (see Figures 7.1 and 7.2), with team members having different cultural backgrounds and professional experiences.

Despite the forming of PDG, XinZin's overall document development process

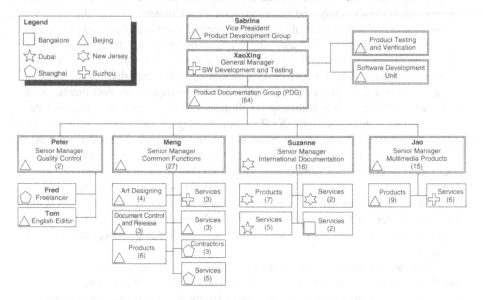

Figure 7.1 XinZin's PDG initial setup. Numbers in parentheses represent the number of people in a particular group.

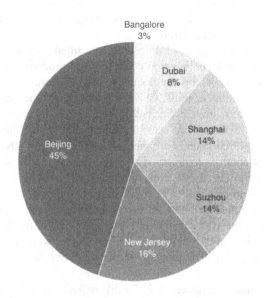

Figure 7.2 PDG global resource distribution. Team members are located across the globe.

remained fragile and documents were being created in all forms and at all quality levels. Xaoxing had done his best in setting the foundation of PDG; however, managing PDG in the long term was never his goal. In spite of his deep managerial and technical background, Xaoxing found it difficult to manage the unknown territory of documentation along with his other core duties.

"Sabrina, I am afraid that the PDG setup in its current form may not sustain for long," Xaoxing raised the issue in a department meeting. "The Chinese documentation team lacked technical writing experience and very few of them are willing to take technical writing as their long-term career. Recently, some of them approached me to discuss the possibility of returning to their previous developer jobs, which they think more promising, exciting, and lucrative."

Sabrina agreed that before things got out of control they needed to bring in someone with a background in documentation practices to take charge of PDG. Sabrina contacted Sanjeev, who was working as a senior documentation manager with a leading telecommunication operator in the United States. Sanjeev, a professional in his late 30s, had a background in computer science, had worked in multiple domains before moving to the telecommunication sector, and had much experience implementing documentation processes. After a few days of consideration and discussions, Sanjeev accepted the assignment. Sanjeev had only about four weeks to prepare for the move, which left him with little time to explore much about China.

FIRST ENCOUNTER

Despite being very tired from a long, sleepless flight, Sanjeev was excited by the opportunity to observe Chinese culture closely, which to him had so far remained synonymous with martial art movies, the dragon dance, and the Great Wall of China. Excitement was still ripe in the air as Sanjeev walked in to meet Xaoxing for the first time.

A confident smile beamed on Sanjeev's face as he extended his hand toward Xaoxing. Without any expression on his face, Xaoxing grasped Sanjeev's hand lightly, bowed awkwardly, completed the handshake quickly, and then moved toward the drawing board.

Speaking English with a Chinese accent that was hard for Sanjeev to understand, Xaoxing gave a brief history of PDG and went on to explain its operational aspects. Xaoxing said many warm words to welcome Sanjeev, but his constant avoidance of direct eye contact and his dull facial expressions were in direct conflict with what he was saying. Xaoxing's English grammar was nearly perfect, but he kept repeating himself and his voice was loud and seemed like shouting to Sanjeev. He appeared to be in a rush to conclude the meeting and did not pause to ask if Sanjeev was following him. By the end of the presentation, Sanjeev was completely exhausted and frustrated over not being able to pick up most of Xaoxing's presentation.

As Sanjeev was walking toward his office, the smell of fresh, ground coffee entered his nose. He closed his eyes, took a deep breath, and allowed the smell to reach his lungs. A little energy seeped back into his body. He quickly reached the cafeteria nearby and with wide eyes tried to read the handwritten menu hanging from the wall of the cafeteria, written in Mandarin. He looked around helplessly but found no one nearby to help him.

"It is easy to identify foreigners in XinZin, particularly those who are new to this place," Sanjeev heard a deep, laughing voice in a U.S. accent. He turned to see the faces of two strangers standing and smiling behind him. They introduced themselves as Tom and Peter. Both helped Sanjeev in locating the item he was looking for and shared their tips on how to place an order in Mandarin. Finally, when Sanjeev murmured the newly learned Mandarin words and successfully placed his order of black coffee with half-sugar, he felt the excitement of experimenting with a new culture.

Sanjeev's smile was back on his face. In contrast to his meeting with Xaoxing that turned sour in a short order, Sanjeev was enjoying the discussion with Peter and Tom. However, things took a turn when he introduced himself as a new member of PDG.

"I am not aware if anyone is joining my team this month. By any chance, are you joining Meng's or Jao's team?" Peter asked Sanjeev, looking puzzled.

"I am reporting to Xaoxing as Director of Product Documentation Group." Sanjeev said politely, sipping his coffee.

Peter's jaw dropped and his eyes bulged right out. "XinZin sure does have its own weird way of managing things. PDG has a new owner and no one in the department is aware of this! When was this decision made?"

Sanjeev was surprised too by Peter's reaction. Sabrina was aware of Sanjeev's joining PDG well in advance, and she must have had her own reasons for not sharing this information with the rest of the group, or with Peter and Tom in particular. Sanjeev chose to remain quiet rather than making any judgments that could prove disastrous for him later on.

"Well, you see, I am still a stranger to this place," Sanjeev tried to hide his surprise. "I am still figuring out how to find my way around here."

There was no further discussion after that. Sanjeev finished drinking his coffee, said goodbye to Tom and Peter, and started walking toward his new office.

"Xaoxing has a bad habit of gathering clowns around him," Tom whispered as he watched Sanjeev leave the cafeteria.

"Only this time, we have to bow in front of this clown," Peter concluded.

After reaching his office, Sanjeev closed the door and slid back on his chair reflecting on the things he had been facing since morning. His first meeting with Xaoxing was disappointing and his interaction with Peter and Tom was not encouraging either. Fully confused with what had been happening to him since morning, he was anxious about what he might encounter in the upcoming department meeting in the afternoon.

As soon as Sanjeev entered the meeting room, he found himself surrounded by a large number of faces that seemed half-asleep to him. The silence prevailing in the room gave Sanjeev jitters. The memories of his first meeting with Xaoxing were still fresh in his mind and the thought of hearing Xaoxing speaking again was painful. Xaoxing briefly introduced Sanjeev to Meng, Jao, Peter, and Tom before proceeding to deliver his welcome note.

In English, Xaoxing gave a brief history of XinZin and PDG. Most of his speech was similar to the presentation he made to Sanjeev earlier. Before Sanjeev could adjust to Xaoxing's voice, Xaoxing switched language and continued his speech in Mandarin.

With little knowledge of Mandarin, Sanjeev helplessly glanced at the other audience and noticed some similar frustration. Peter's face was wrapped in layers of turmoil and disappointment. Tom was trying to project a calm image, but he also had a perplexed look on his face. Even the Chinese, Jao and Meng, were sitting like statues—quiet and motionless.

Why can't he speak in Chinese and have an interpreter to translate in English? Sanjeev thought. During the entire speech, Xaoxing frequently smiled and looked at Sanjeev as if introducing him and mispronounced his name as "*Sen Jei.*" At one point during the speech, Xaoxing smiled at Sanjeev and paused slightly longer. Sanjeev thought that was Xaoxing's signal for him to address the PDG members. But as he was preparing to stand up, Xaoxing again turned his face toward the audience and continued with his speech. Sanjeev quietly rested on his chair, trying to hide his embarrassment.

When Xaoxing finally finished his speech, he moved toward Sanjeev and invited him to address the PDG members. By now, Sanjeev was left with little interest to introduce himself. With a fake smile on his face, he started half-heartedly, trying to keep the nervousness out of his voice. He talked about his vision for the potential and future directions of PDG, which was met by a complete lack of enthusiasm from the audience. Thinking that he was being cued to finish up, Sanjeev closed his speech earlier than planned and quietly moved toward his chair. His speech was followed by few dull claps and then complete silence.

Sanjeev then called on a few individuals to get their ideas about how the things for which they were responsible would fit into his vision for PDG. Peter and Tom came up with some mixed opinions about why certain things might or might not fit well. On the other hand, Chinese team members mostly remained quiet or answered vaguely. When Sanjeev pushed them to get specific information, a couple of them simply said, "It's ok." Sensing that the meeting was not achieving his objectives, Sanjeev thanked everyone for coming and ended the meeting.

Walking out of the meeting hall, Sanjeev could not help feeling disappointed about his new boss, Xaoxing. In the two meetings Sanjeev had with him, he talked about useless issues in the industry when more pertinent department information had to be discussed and explained. Sanjeev was disappointed that Xaoxing's approach had taken away from him an excellent opportunity to break the ice with PDG members.

After the disappointing first day at XinZin, Sanjeev's head was bursting with more questions than ever. Later that night, crawling under the covers of his bed, Sanjeev tried to reflect what went wrong today and why he failed to catch people's attention on the first day. After all, it never happened to him even with some of the most aggressive clients he had had in the past, but this time things were radically different and he had no clue why.

Maybe tomorrow will be a better day for me, he told himself before drifting off to sleep.

PANDORA'S BOX

When Sanjeev met Xaoxing the next day, he was full of questions about the behavior of PDG members he observed yesterday and about Sabrina's decision to appoint him instead of trusting any of the existing managers to manage PDG.

"Unfortunately, of late, none of the senior managers are worth their salt," Xaoxing explained. "If they were capable of making any changes out here, they could have achieved their targets by now. Years of their association with XinZin have made them complacent. They can understand the problems, but take no efforts to find a solution. Instead, they are busy pointing fingers at others for failures."

Sanjeev's eyes were wide open and he was all ears waiting for some great truth to come from Xaoxing's lips. Xaoxing paused, leaning back on his chair and recollecting his thoughts, "I have always had high opinions about Peter, but his approach is too simple and ineffective to manage the complex problems we are facing. Suzanne is least bothered about what happens in PDG China. Her interaction with others is confined to project-related activities in the United States. Meng and Jao still need guidance at every step to deal with international customers and are still reluctant to socialize with Suzanne's team and the English editing group, even if I push them hard."

Xaoxing was speaking in a natural pace with minimum hesitation or repetition and his pronunciation was much easier to follow. His facial expressions and body language seemed more in tune with his words. Unlike yesterday, Xaoxing was projecting an image of someone well organized, honest, and in total control of the things around him.

"But, I think no one in the PDG was prepared for such changes in the department leadership," Sanjeev shared his concerns. "Not many team members were comfortable in the meeting yesterday and I presume that my speech was not pleasing to all ears."

Xaoxing laughed heartily, shook his head in disagreement, and said, "That's because, except for a few, none of them understand English."

Sanjeev was stunned. Neither Sabrina nor anyone from human resources at any stage of discussion had ever told him that he had to be fluent in Mandarin. Sanjeev sat on the chair with his head down and did not look up for a while. How did I get myself into such a messy situation? He thought. It never crossed his mind before accepting Sabrina's offer that language problems could be such a problem. He sensed a spike of panic.

With a worried look on his face, he asked Xaoxing, "Is the China group developing documents in Mandarin only?"

"Mostly, but not all," Xaoxing replied. "For products developed by China R&D, the PDG China group produces the documents in Mandarin. If any international customer requires the product, Meng's team translates the documents and hands them over to Tom for language editing and polishing. Similarly, for products developed by U.S. R&D, Suzanne's team delivers the documents in English and Meng or Jao takes the ownership to translate the documents into Mandarin."

Sanjeev felt more restless upon hearing this. The world was not as flat as he thought. Without having any expertise in Mandarin, he was clearly poles apart from the rest of the team members, and chances were high that this gap was going to widen as he started interacting with them more often. He was walking on the edge of a razor blade in his bare feet. With one last look at Xaoxing, he left the room.

Mandarin was undoubtedly the official language for communication at XinZin with no exceptions for foreigners. Sanjeev started feeling the heat when dozens of e-mails written in Mandarin started reaching his mailbox. Without knowing fully what lay hidden in those unfamiliar characters, it was difficult for him to understand the issue or take proactive actions. Even the information on the company's intranet was available only in Mandarin. Sanjeev soon found himself struggling to perform even basic tasks on his own. Overwhelmed with the language problem, he called upon his department assistant to help him understand the intranet interface and translate e-mail content. The arrangement worked for a while, but now Sanjeev had another problem. Depending too much on his department assistant had the possibility of un-knowingly exposing some confidential data to her. In a move to become more inde-pendent in his work, he installed some free Mandarin-to-English translation tool but soon got frustrated with the poor quality of translations coming out of the tool.

The language problem got out of control when Sanjeev locked horns with the manufacturing team over some critical document releases. The manufacturing team needed Sanjeev to approve the release of certain documents to the field. Having sent some repeated reminders and never hearing from Sanjeev, the manufacturing team escalated the matter to Xaoxing and Sabrina. Sanjeev had no idea what had caused the escalation. He was responding to e-mails regularly and did not remember receiv-ing any approval request mails from manufacturing. He again scanned through all the mails, including spams that he had received so far, but still failed to locate any approval request mails. Confused and frustrated, he approached Xaoxing.

That was when Sanjeev received another shock: The reminder mails were in fact system generated and were lying idle on the intranet portal for a while. Since the en-tire content on the intranet was in Mandarin, Sanjeev had no idea that something as important as release approval requests could be hiding under the layer of foreign characters. It took Sanjeev another hour to approve the release and to resolve the is-sue, but the episode left him more disappointed with the PDG setup. No one in the department, including Xaoxing or his department assistant, ever bothered to tell him about such triggers happening through the intranet, particularly when both of them were aware that Sanjeev was blind and deaf to Mandarin. As a result, a task that could have been completed in an hour took weeks.

In his frustration, Sanjeev accused his department assistant for not telling him about the intranet approval process and causing a near fiasco. His department assis-tant, in turn, found Sanjeev's behavior rude and impolite and approached Sabrina. Sabrina spent the better half of her day trying to convince both of them that it was simply a case of miscommunication and the entire episode should better be ignored. The incident alerted many people of Sanjeev's poor Mandarin skills. However, most of them were reluctant to communicate with him in English because of their lack of self-confidence or skill in English. Therefore, they identified Sanjeev's assistant as their single point of contact to relay any matter to Sanjeev.

There seemed no end to the language woes. Sanjeev was dismayed at the quality of English documents written directly or translated from their Mandarin version by the Chinese group. The documents were full of grammatical errors, missing or incomplete information, and procedures that could rarely help anyone achieve anything. When, on one occasion, he located a few such documents that had been recently shipped to some international customers, he feared that the number of errors in other documents could be even higher.

When asked about procedures being followed to locate and fix such quality issues, Jao and Meng had no clear answers, and Peter dodged the question by shifting the focus back on Jao and Meng. "Sanjeev, we are editors and my group only takes care of editing the content that Meng and Jao's writing teams provide. Clearly, we don't have any control over the content."

Upon hearing Peter, Meng began speaking in a high-pitched, breathless voice, allowing her words to trip over themselves, "Sanjeev, I think this is an isolated case and should not be used to judge the quality of all the deliverables that we produce. In fact, ever since we shipped these documents to the customer we have not received any complaints from them. Our writers are more focused on writing than editing the content, and if there are any language issues, it is the English editors' responsibility to fix them."

Before Meng could complete her last sentence, Tom nearly shouted, "Talking about responsibilities! Meng, for your information, the editing team is investing additional time and effort to ensure that the documentation set works as a whole. Our writers are mostly inexperienced; they do not realize that an installation manual, for example, should be written in a different style than an operator's manual, which in turn is targeted to a different user than a system overview. They also sometimes have difficulty determining which information belongs in which manual. In fact, we are often called upon to rewrite manuals, assembly instructions, and directions. The major problem with the original, aside from grammatical and spelling errors, is with the document content. There isn't enough information to do a task properly!"

Jao shouted back at Tom, "The problem with you is you do not bother to know what others are asking. Your focus is to insult others and impose your answers on them. We know our task better than you do. Just because you speak flawless English, it doesn't give you the right to insult others nonstop. Despite being called rude for years, you have not changed at all. How many people in XinZin had called you rude and asked you to shut up? Not a while ago, someone told me that you made racist comments to him. You have a history of rotten behavior. No one can stop you. Do not, however, think people will not fight back."

Tom was equally infuriated. "Perhaps this is a case of feeling threatened when someone points out something we feel we should have done but didn't. This is because people not directly involved with documentation development typically underestimate the complexity of producing a quality document and don't give much thought about the document accuracy, consistency, and usability in terms of cost."

Tom now turned to Sanjeev, "Sanjeev, often the condition of the documents we review is so horrendous it takes four or five editing cycles before it is even marginally acceptable. More often, the document just ceases to return to us after one or more revisions. We don't usually get to a point where the editors are satisfied with the quality of the document before it is published. The writers don't even refer to the style guidelines while drafting the documents. This is a problem affecting all Chinese writers. It's not anyone's fault; it's just the way things have been done in PDG for years."

Peter nodded at what Tom said. "Sanjeev, I think you can easily focus on our group because we all speak English—but significant improvements will only derive from focusing on the feeder documents to our group: the output of the technical writers. Many of our edits were repeated numerous times in each of these reviews, and later we find the same problems in the new document. I am not surprised if these errors are also present in the documents written in Mandarin. Just because we could not read them does not mean that they are free from errors. "

Meng intervened, "My writers are often confused about UI description and style guidelines. For instance, what to call that little box that pops up to tell you a procedure has completed successfully. Whenever we ask English editors, they refer us to the style guide that they have prepared. They do not help us when asked and later on complain that we don't share our concerns with them."

A sarcastic smile passed over Tom's face. "I am not sure what the problem is. Maybe the writers just don't want to hear anything from us. In the past, we have received several requests from them not to find so many errors in their documents because it embarrasses them. The problem is people here don't like you to find problems with their work. They would rather risk the ire of a seemingly distant customer than lose face in the group."

Clearly, there was nothing sensible happening in the room. The points of discussion were sidelined and PDG managers were engaged in a war with each other. Before Sanjeev could realize, they started calling each other names. In the process of blaming each other, they did not even care what negative impressions they were making on their new boss.

In addition to the conflicts between Chinese writers and English editors, Suzanne's team was struggling with their own problem working with Chinese translators in Meng's and Jao's teams. In one of the discussions with Sanjeev, Suzanne vented her frustration, "Sanjeev, I have been telling my fellow Chinese writers that good translation is not as simple as translating word to word; the context needs to be considered as well! Most often, the translators add or frame sentences in a highly formal language that sounds ridiculously old-fashioned. For instance, there is no need to start every step with 'please.' Also, 'Click the button' sounds much better than 'Click the button hardly' or 'Click the button softly.'"

Sanjeev took a deep breath and Suzanne continued. "I remember one of the parameter descriptions in the English version of the user guide was 'Specifies the network adapter for which to query or set parameters.' It goes in Mandarin as '为了设置参数，您询问或者指定网络适配器,' which in English means, 'In order to set the parameter, you inquire or designate the network adapter.' My all-time favorite is 'Click OK to continue,' which translates to '继续的好单击.' In English, this means

'Continuous good one-click.' Nobody on the Chinese team wants to know that a term in English may not translate word for word into a Chinese equivalent because the English word has different meanings in different contexts, some of which do not exist in Chinese. Similarly, English cannot completely explain many existing concepts in Mandarin because of their subtle and intricate connotations. An otherwise correct term in Mandarin presents a completely different interpretation in English when it is translated word for word. This leads to confusion, doubt, and sometimes embarrassing situations in the field."

Suzanne continued: "Last month, one of our customers nearly lost the equipment due to an erroneous statement mentioned in our documents: 'Normally equipment temperature is about 10–20 degrees Celsius (50–68 degrees Fahrenheit) higher than the equipment room (ambient) temperature.' As we know, an increase of 10–20 degrees Celsius is the same as an increase of 18–36 degrees Fahrenheit, and not 50–68. Since we are talking about a relative temperature increase, we need to use the formula $(C \times 9)/5$, and NOT $(C \times 9)/5 + 32$, as you would when calculating absolute temperature. This is clearly not a language issue but a logical one. I've come across such mistakes several times, but none of our Chinese writers are sensitive to this issue. We need translators who know both Mandarin and English but also are well trained in the technical substance of what they translate. Such people are very hard to find and training them takes a long time."

MORE THAN LANGUAGE

As he started getting deeper into PDG operations, Sanjeev realized that the communication problem was even more deeply rooted than the language barrier. There was not any centralized location to create, track, and manage documentation projects and resources. Project documentation within PDG was mostly ad hoc and confined to weekly status reports giving only the list of deliverables on which each writer was working. The reports lacked information on the expected and actual start or finish dates of document deliverables, their current state, how they were related to a project, or any issues impacting the delivery of the documents. After spending weeks reading these status reports, Sanjeev could only find out who was assigned to what product lines and conclude that the documentation practices at XinZin were mostly rudimentary. PDG had no way of knowing what products were being shipped to the customers and when. Without proper planning, PDG managers were not aware how long the writers should wait until the development effort was farther along before starting to write or how an additional feature might affect document development efforts.

In an attempt to streamline project management activities, Sanjeev made a new request of his managers. "Going forward, in weekly meetings please share a status summary of all projects that your team members are working on. Also, let's plan to have your group's operations reviewed on a monthly basis."

Sanjeev's announcement caused ripples among PDG managers. Jao could not suppress his quirky smile, "It would be nice to have some sort of comprehensive project plan, but I know this is far from being practical. Getting the information you are looking for is time consuming and requires us to spend more of our already limited time."

"But without any plan in place, how can we be sure that we have achieved our goals?" Sanjeev asked. "If it takes some extra effort to collect and present the information, I am fine as long as you provide me enough visibility on the projects your team is handling."

While Peter, Tom, and Suzanne provided Sanjeev with a detailed review of their team, projects, and issues at hand, discussion with Jao and Meng turned out to be a frustrating experience. In spite of Sanjeev trying to give clear guidelines on what project aspects he would like to know, Jao and Meng did not seem able to give Sanjeev straightforward answers. Instead, in their reviews, they talked about their current positions in the organization, the company's future direction, and so on.

Sanjeev rescheduled the meeting with Jao and Meng at least three times to give them extra time to prepare their presentation. He also organized discussions with Jao and Meng to help them understand what he needed from them. Jao and Meng sat silently when Sanjeev talked. They seemed more interested in watching the way Sanjeev's hands moved than listening to what he had to say. At the end of the discussion, Sanjeev asked Jao and Meng if they got the instructions. Both of them remained silent and Sanjeev took their silence as affirmation.

But once again, Jao and Meng ignored Sanjeev's instructions and presented some half-complete and poorly organized data. Sanjeev's patience was growing thin as Jao and Meng talked. Finally, he intervened in an angry voice, "All I want to know is your project's status, but why can't I get a simple answer?" He left the room, banging the door behind him.

Meng sat quietly in the meeting room staring down at the floor. She thought she was doing just as Sanjeev instructed her to do, yet she was always losing face in front of her colleagues and boss. She had no idea how to satisfy Sanjeev.

"Is he going to fire me now?" She asked with fear and started weeping bitterly. Infuriated by the embarrassment and disturbed by Meng's tears, Jao did something he had never done before. He walked straight into Xaoxing's office and narrated the entire episode. "I cannot rest in this uncertainty," he said fuming with anger. "How come this foreign director is so aggressive and rude to us?"

"I didn't know Sanjeev could get so angry," Xaoxing said, although he was not totally surprised about the episode. He had sensed that something was wrong. Instead of solving issues and making any difference, Sanjeev seemed to have caused only conflict. According to Jao, Sanjeev was projecting himself as someone in charge of PDG, and he was putting his nose where it was not wanted. Jao complained that Sanjeev forced PDG writers and team leaders to participate in weekly project meetings even though most of them were uncomfortable communicating in English; he embarrassed them further by locating errors in the documents and asking for correction; he questioned whether writers know "their stuff" in what they were writing.

Xaoxing decided to confront Sanjeev. When they met, Sanjeev tried to defend his actions, "I have no idea why Jao and Meng cannot prepare a simple project status report with the level of detail I need. I have provided them sufficient information over the last few weeks and I don't see any reason why the task cannot be completed on time to my satisfaction."

Xaoxing didn't say anything; he looked up at Sanjeev and stared in his eyes. Sanjeev was standing motionless, with anger, sadness, and loneliness screaming from his eyes.

"Stop locating fault with everyone," Xaoxing said calmly, patting Sanjeev on his back. "Just give some time to Meng. I am sure you will get along with her better."

Xaoxing then tried to explain Jao's viewpoint to Sanjeev, and Sanjeev's anger flared again. Sanjeev could see clearly that the managers were on the wrong track, but instead of correcting them, Xaoxing was trying to cover up their acts. Perhaps Xaoxing's proximity with Jao and Meng had made him blind to their acts.

Frustrated, Sanjeev recounted to Xaoxing a recent encounter with one of the customers. For the past few months, the client had been approaching every possible level of authority in XinZin to get supporting documents for his site, but the issue remained unresolved. When the matter was brought to Sanjeev's notice, he found that the reason not much progress was being made with the client was that none of his managers, including himself, knew enough about the client's requirement to propose a solution. He called in both Jao and Meng and asked them to develop a detailed report on the client's requirement.

Two days before the report was due, Sanjeev asked Jao how it was coming. Jao said that everything was going as planned. However, a few hours before the report was due, Meng helplessly told Jao that the report would not be ready. Meng explained that even though she had worked into the night every night to complete the report, it was just impossible and she had doubted from the beginning whether she could complete the report in a week. Sanjeev was angry and did not understand why Meng had said nothing in the first place. He headed straight to Meng's desk, yelled at her, and left her in tears. The delayed report made it hard for Sanjeev to convince the client that a proposal would be ready in another week's time.

Xaoxing listened to the incident in silence and let Sanjeev continue. "Have I done something wrong by asking Jao and Meng about the project status? Is it my fault if they cannot follow simple instructions? I think not. I have done my best to help them, but I cannot help if they are not willing to leave their ego out of the equation and instead focus on the task. You don't even know what I am going through," Sanjeev told Xaoxing blatantly. "If Meng and Jao have any concerns or they have difficulties understanding my instructions, they should talk to me, rather than escalating the problem to you or anyone else for that matter."

Xaoxing remained quiet. He thought that Sanjeev's prejudice about people around him had grown to the extent of making him rude. To him, Sanjeev seemed egoistic. If he was so superior, why was he not able to get along with his team, especially Jao and Meng? If he was so superior, he should have easily identified and resolved the issues in the first place, so why is he still struggling? Instead of improving the situation, Sanjeev was adding fuel to an already raging fire, trying to force his ideas on everyone and take over everything. Constant rifts between him and PDG managers had transformed PDG into a "compartmentalized" terrain and team morale was at an all-time low.

Sanjeev had so far failed to strike rapport with Chinese writers and as a result they were more comfortable communicating their concerns to Jao, Meng, and Xaoxing rather than to Sanjeev. Earlier on, they had wanted to discuss documentation issues with Sanjeev and to learn from his experience; however, now they were distancing themselves from Sanjeev after noticing his "annoying habit" of finding too many problems with their work. The entire Chinese team, particularly Jao and Meng, was feeling isolated after noticing Sanjeev's inclination toward Peter and Tom. He was hanging out with Peter and Tom more often than with the entire team combined together.

But in spite of the confrontations between Sanjeev and the Chinese group, Xaoxing also noticed that international customers were more comfortable discussing their issues with Sanjeev than with anyone in the group, including himself. Sanjeev was the only one who was listening to their concerns in length, communicating with them more frequently than anyone else, and suggesting alternatives to resolve their issues. Customers were showing signs of improved confidence with PDG. In some critical cases, Sanjeev had been able to negotiate a win–win deal with the client. He remembered how in one instance Sanjeev used some existing documents to draft, within few days, a document that was able to address the customer's requirements. None of the PDG managers were able to do such things. Maybe if Sanjeev could make himself comfortable with the group, he could do wonders for PDG.

Sanjeev leaving at this stage could spell doomsday for the PDG improvement plans, Xaoxing concluded. But something must be done, fast!

WAR WITHIN

With each passing day, Sanjeev's frustration was growing. There were many incidents, big and small, that had made his last few months frustrating, but Sanjeev was too tired to remember them all. He was at war with everyone around him, but more than that, he was at war with himself. He thought that working with Chinese team members was like working with people from another planet. He just couldn't communicate with them, and he could never figure out what they were thinking. It was driving him crazy. He was getting tired of XinZin's working conditions, tired of not understanding anybody or being understood, tired of not getting the information he wanted from his team members and colleagues, tired of being a passive attendee at department or group meetings, and tired of not being able to analyze issues or solve them. His feeling of isolation had grown to the extent that he was now getting scared of talking to people even on issues that were once close to his heart. He was starting to question things that he had never questioned before. Something must obviously be wrong with me, he thought. Or, was XinZin the wrong choice for me?

Sanjeev remembered what Fred, the freelance writer from the English editing group, once told him, "If you ask about mutual trust, I would say that it exists within the team but not across the teams. Chinese team members share mutual trust and are entirely tolerant of mistakes made by each other. These values often take a U-turn in case of foreign employees, who have no room to make any mistakes."

Xaoxing had so far led Sanjeev to believe that this was just another assignment, but Sanjeev lost all hopes of turning the situation around in his favor. For him, Xaoxing became an authoritative boss who seldom valued his opinion. He was against Sanjeev's working style and seemed to see his way of trying to solve problems as fault finding. He was of the view that if Sanjeev had to bring about any change, he had to follow the line of thought which had been well understood and accepted in XinZin.

Sanjeev felt that he was going through one of the worst periods of his career. His lack of Mandarin skills was overshadowing his managerial and technical skills. He had never had problems communicating with people, but today he could not even ask for a document from his team members without needing an interpreter to translate his words. It was painful for him to keep a fake smile on his face while watching others speaking to each other and laughing together near coffee machines or in a department gathering. During department meetings, on countless occasions he pointed out to his colleagues that he could not follow conversations in Mandarin and therefore would appreciate the conversation in English, but his request was ignored every time. To his embarrassment, the minutes of meetings were also written in Mandarin and every week he had to spend extra time with his department assistant to understand them.

Not too long ago, frustrated with the frequent lack of communication with the Chinese team, the Dubai team approached Suzanne to facilitate the discussion on their behalf with the Chinese team as the communication gap between them was too wide to bridge. Sanjeev spent extra time with Jao, Meng, and Suzanne discussing how to close this gap. As Sanjeev concluded the discussion, he realized, surprisingly, that Jao and Meng were equally disturbed by the communication gaps and they seemed to be doing the best they could to help everyone. It was only because of their team members' poor English skills that they were not able to present their viewpoints in the right manner and often ended up being the culprit and victim at the same time. So it seemed I wasn't the only one banging his head against a wall that has not even budged an inch, thought Sanjeev. Perhaps, it wasn't anybody's fault, and yet no one had any idea on how to break the deadlock.

It was already dark outside and lights were turning on along the street. Most of the people had left the office and the traffic outside had started to ease. From his office window, Sanjeev could see the beautiful campus and the roads that stretched beyond the building. A few people were walking about but mostly it was quiet. Sanjeev felt the loneliness in the air. He looked out the window wishing that somehow he could do something to fix everything, or turn back things, or something. He was unsure how to approach this. He was sorry for many things, but mostly he was sorry for making himself into a crude joke. He had not anticipated what was before him when he joined XinZin and he was now finding out that being a foreign manager in a Chinese company was unbearable. He was desperate to leave XinZin and go back to where he belonged—somewhere where he could use his skills for a better cause.

Managers don't know what their writers are doing, writers don't know what projects they are working on, and I don't know what all of them are up to. It can't get worse than that! The exhausted and despairing Sanjeev thought. Enough is enough. I can't stand it anymore.

COUNTDOWN

Next morning, Sanjeev met Xaoxing to convey his decision. "Xaoxing, I have done what I can. Unfortunately, things are not improving and people here have no consideration for me," Sanjeev said in a low tone. "Before making matters worse, I think I should leave XinZin."

Xaoxing didn't seem too surprised. Taking a sip of his coffee, he said, "Sanjeev, it's not fair to put all the blame on other people when they don't respond the way you want them to. The blame has to be put squarely on your shoulders too. As team owners, we need to set our people up for success, not failures. Once you take that responsibility, once you decide that it is your job to think like them, then and only then, will you get their support."

Sanjeev sat silently with a blank look on his face. He had no answer to what Xaoxing said.

"Now that the number of foreign employees in XinZin is growing, management has taken some initiatives to ensure better teamwork," Xaoxing said. "I had been thinking that maybe this is what you and what PDG need. Why don't you meet me tomorrow and we will talk more about this. "

"All right. If you think anything can be done at this stage, I am open to give it a try." Sanjeev said in a disappointed tone. He was not enthusiastic about experimenting with this assignment any further.

Next day, when Sanjeev entered Xaoxing's room, he found him surrounded by Meng, Jao, Peter, Tom, and a stranger.

"We were waiting for you," Xaoxing flashed a smiling look at Sanjeev. "Meet Aaron, who recently joined XinZin as a training consultant. He is working on a cross-cultural training proposal that you may find interesting. This training program has been in the pipeline and has been recently approved by management."

Aaron's proposed cross-cultural training had components related to dealing with general cultural dimensions, national values, and workplace incidents. The program goal was to make participants sensitive about cross-cultural issues and provide a platform where people could talk openly about the issues they face everyday in cross-cultural contexts and analyze the problems in different cultural contexts. The training first provided trainees the knowledge about national cultures and attitudes in a host country and then helped trainees to handle the situations in simulated cross-cultural environments. The training included language studies and skills training through case studies, simulations, and behavior modeling.

Within a few days, all PDG managers, including Sanjeev, had multiple meetings with Aaron to talk about the kind of problems they were facing and how they perceived those problems. Each member also shared what they would like to learn in a training program and the communication area on which they would like to focus.

A few days later, Peter entered Sanjeev's office with a big smile on his face, "You won't believe what I just heard from Aaron. He told me that he was soon going to start the cross-cultural training for us, including Mandarin training. In my three years at PDG, it's the first time Mandarin training has been offered to us. Would you like to join us?"

Sanjeev's face beamed with excitement, "Of course, I would like to join."

All the Chinese and foreign managers working with XinZin's R&D teams attended the program. The program helped them compare and contrast Chinese business styles with business styles in other countries and gave them an opportunity to learn to communicate with each other more effectively. During one of the simulated exercises, Aaron asked Sanjeev to present the project status report to his Chinese boss. Aaron wrote a few important points about the report on a piece of paper. He then translated the text into Chinese, instructed Sanjeev on how to pronounce the text, and asked him to present the information to one of his colleagues who was acting as his boss.

At first, Sanjeev thought it was a child's-play, but as soon as he started speaking, his voice tumbled out of his mouth awkwardly. It didn't sound right to him. He tried many times but failed repeatedly. He could see that his "boss" was looking angrily at him, waiting for him to pass the information, and out of nowhere Sanjeev started sweating like crazy.

Aaron asked the person playing Sanjeev's boss, "What do you think is the problem with Sanjeev? Can you help him understand how he can improve?"

"Maybe he is too shy to speak in front of the audience. Also, I think he is not following the instructions given to him," the "boss" said, "but I don't see any problems with the instructions. They are simple and easy to understand. So, maybe he should focus on improving his public speaking skills to get better results next time."

"Can you see now how difficult it is even for an experienced speaker to communicate in a foreign language?" Aaron turned to Sanjeev and the other trainees. "The other person may easily misinterpret your behavior and the entire situation and soon things may spiral out of control."

Sanjeev now realized how he had made the life of people around him difficult with his misplaced observations and terse statements. He lost count of how many times he had behaved rudely with Jao and Meng, without realizing how insulted or indignant they must have felt.

The few weeks of attending the cross-cultural training program had developed a new level of understanding among PDG managers. They could relate the role-play exercises to the situations they had faced earlier and realized how their ignorance had made them interpret the situations the wrong way. Every training session was opening a new window of cultural awareness and giving them a new perspective of the people around them. The managers were learning to understand the problems with their own behaviors, to be more forgiving of others' mistakes, and to correct those mistakes without offending people. The training helped everyone realize that people often have the desire to improve, but they sometimes feel helpless while dealing with foreigners, particularly in a foreign language and environment.

After getting overwhelmingly positive feedback from participants, Aaron extended the program to other XinZin groups, such as marketing, sales, and supply chain. Aaron worked overtime with Sanjeev, Meng, and Peter to improve the program and extended it to other business units in Suzhou and Dubai and to some of the XinZin customers in Russia and Japan.

With all of this work, things started improving in PDG. Although its documentation system continued to remain rudimentary for some time, there were positive signs and people were starting to embrace the new reality and contribute what they could to improve things.

In one of the weekly meetings, Jao placed a request, "Sanjeev, can you please ask Suzanne to give us some training on Framemaker? The templates and scripts they use can save us a lot of time in developing the document."

This was the first time in Sanjeev's stay with XinZin that Chinese managers were coming up with such a clear request. He responded swiftly, "Please check directly with Suzanne. I am sure you two will find a way to cooperate on this."

Suzanne's response was equally positive. "We interact with Jao's team very frequently," she told Sanjeev, "and we know that they have a tough time applying consistent styles, formatting, and layout to the Framemaker documents. I am sure with this training some of their tool-related issues can be addressed, which should improve their confidence with the tool."

On another occasion, Meng and Jao visited Sanjeev's office and handed him a handwritten sheet of paper, "We came to know from your assistant that you have difficulty understanding workflows in Mandarin. This list can help you find your way."

Sanjeev glanced at the sheet. Written on one side of the sheet was the list of the workflow keywords in Mandarin and on the other their corresponding meaning in English. He was elated. With this list, he could now navigate through the intranet more independently. For a few seconds, his face beamed with joy and then immediately filled with sadness when he remembered how badly he had behaved with Meng, Jao, and others in the past.

"I am sorry for my past behavior and for all of my actions that have offended you," Sanjeev said after a long pause. "The mistakes were all mine. You wouldn't have had to face the trauma if I hadn't tried to change you into something you're not." Sanjeev said with an apologetic face. "That reminds me, I have to apologize to the entire group."

"You don't have to," Meng smiled. "You are no different than us. If you were mistaken, we were too. The good thing is we realized it before it's too late."

"Thanks, I mean *Xiè xiè,*" Sanjeev said.

"*Bú xiè,* I mean don't mention it," Jao said, and everyone laughed.

"I also have something that you have been asking for," Jao said. "I have placed the current status of all the projects in a shared location. Meng and I have taken the responsibility to update it frequently so that you can see how things are moving and correct us in plenty of time."

With help from Jao and Meng, Sanjeev started interacting with clients more frequently and helped the Chinese group understand the clients' requirements and respond with the right solution. The Chinese team, on the other hand, committed themselves to turn Sanjeev's plans into reality and the results were encouraging as old and existing issues with the customers were being solved.

The weekly discussions also became more relaxed and people opened up to listen to other's concerns and acted on them. For instance, after getting to know the problems Suzanne and the Dubai team were facing while communicating with Jao and Meng long distance, Peter volunteered to facilitate the discussion between the two groups and helped to solve some critical issues. Meng and Jao returned the favor by helping English editors direct technical queries to development teams. Conducting a face-to-face discussion with Chinese developers, which had been near impossible for the English editors so far, suddenly became possible. Peter and Tom were surprised to know the depth and vastness of the developers' technical knowledge.

Tom was not left behind. He came up with a list of frequently encountered errors in the documents and arranged a session for the Chinese writers to be aware of these problems. He later developed and distributed quick reference cards, which Chinese members could place on their desks and refer to while developing documents in English. This time, no one felt embarrassed or humiliated. It was fun all around and the tension that existed between English editors and Chinese writers slowly started fading away.

Sanjeev was simply elated and surprised to see the changes happening in PDG. Frequent interaction with the Chinese group also led him to change his earlier perception about the group and helped him to become more approachable and friendlier with them. His Chinese team members were now openly discussing documentation issues that were bothering them and turning to Sanjeev for help. In exchange, Sanjeev was learning the difference between Mandarin and English writing styles and their respective roles in shaping XinZin's technical documentation. He no longer avoided contact with Chinese team members and could surprise them by speaking common sentences in Mandarin or by near-perfect handling of chopsticks. Responding to his positive approach, people around him were now more considerate to him than ever before.

Of course, it was not just smooth sailing from there on. The conflicts between the English editors and the Chinese group did not simply go away and occasionally required Sanjeev's intervention. Sanjeev still struggled to accept some facets of the Chinese belief system. For instance, at XinZin, decisions were always made at the top and never questioned at the lower levels. It seemed that the ancient Confucian principles still provided much of the basis for Chinese organizational bureaucracy, respect for seniority, rituals of etiquette and ceremony, and various types of business relationships. But Sanjeev's more frequent and open interaction with his Chinese counterparts during and after the training gave him a jumpstart to understand these and other aspects of Chinese culture. He now realized that his mission to become a successful manager can succeed or fail according to how well he can understand and accommodate cultural nuances.

"Are you still considering moving out of PDG?" Xaoxing asked Sanjeev one day.

"I don't see myself among aliens anymore," Sanjeev smiled. "I am now working with *my* people."

ACKNOWLEDGEMENT

Thanks to Professor Alexandra Y. Benz-Denaxas (adjunct faculty, Indian Institute of Management Bangalore) for many helpful comments and suggestions on drafts of this chapter.

RECOMMENDED READINGS

1. The kind of cross-cultural training mentioned in the story is variously referred to as intercultural training, diversity training, and multicultural training.

These programs are common and can be successful in the workplace—although we also need to be cautious of their actual effect and limitation. The following readings overview the history, models, techniques, and outcomes of intercultural training:

- W. B. Gudykunst, R. M. Guzley, and M. R. Hammer, "Designing intercultural training," in *Handbook of Intercultural Training,* 2nd ed., D. Landis and R. S. Bhagat, Eds. Thousand Oaks, CA: Sage, 1996, pp. 61–78.

- M. D. Pusch, "Intercultural training in historical perspective," in *Handbook of Intercultural Training,* 3rd ed., D. Landis, J. Bennett, and M. Bennett, Eds. Thousand Oaks, CA: Sage, 2004, pp. 13–36.

- S. M. Fowler and J. M. Blohm, "An analysis of methods for intercultural training," in *Handbook of Intercultural Training,* 3rd ed., D. Landis, J. Bennett, and M. Bennett, Eds. Thousand Oaks, CA: Sage, 2004, pp. 37–84.

2. The training program described in this story dealt with, among other things, "general cultural dimensions" and "national values." These cultural dimensions and national values are often based on the cultural heuristics and models developed by scholars such as Edward T. Hall, Geert Hofstede, and Fons Trompenaars and applied in subsequent works. Consult the following sources for more information on cultural heuristics and models:

- E. T. Hall and M. R. Hall, *Hidden Differences: Doing Business with the Japanese.* New York: Anchor, 1987.

- E. T. Hall and M. R. Hall, *Understanding Cultural Differences: Germans, French and Americans.* Boston: Intercultural, 1990.

- F. Trompenaars and C. Hampden-Turner, *Riding the Waves of Culture: Understanding Diversity in Global Business,* 2nd ed. New York: McGraw-Hill, 1998.

- G. Hofstede, *Cultures and Organizations: Software of the Mind,* 3rd ed. New York: McGraw-Hill, 2010.

- R. R. Gesteland and M. C. Gesteland, *India: Cross-Cultural Business Behavior.* Copenhagen, Denmark: Copenhagen Business School Press, 2010.

- W. Hu and C. L. Grove, *Encountering the Chinese: A Guide for Americans,* 2nd ed. Yarmouth, ME: Intercultural, 1999.

3. Although it is meaningful to learn about cultural heuristics and models as developed by Hall, Hofstede, Trompenaars, and others, it is important to note that these works are not without controversies. The sources below illustrate some of these controversies:

- B. McSweeney, "Hofstede's model of national cultural differences and their consequences: A triumph of faith—a failure of analysis," *Human Relations,* vol. 55, no. 1, pp. 89–118, 2002.

- D. Williamson, "Forward from a critique of Hofstede's model of national culture," *Human Relations,* vol. 55, no. 11, pp. 1373–1395, 2002.

- R. P. Hunsinger, "Culture and cultural identity in intercultural technical communication," *Technical Communication Quarterly,* vol. 15, no. 1, pp. 31–48, 2006.

- P. W. Cardon, "A Critique of Hall's contexting model: A meta-analysis of literature on intercultural business and technical communication," *Journal of Business and Technical Communication,* vol. 22, no. 4, pp. 399–428, 2008.

- P. Signorini, R. Wiesemes, and R. Murphy, "Developing alternative frameworks for exploring intercultural learning: A critique of Hofstede's cultural difference model," *Teaching in Higher Education,* vol. 14, no. 3, pp. 253–264, June 2009.

- T. Fang, "A critique of Hofstede's fifth national culture dimension," *International Journal of Cross Cultural Management,* vol. 3, no. 3, pp. 347–368, 2003.

DISCUSSION QUESTIONS

1. The story contains an example of an erroneous statement concerning the comparison of Celsius to Fahrenheit temperatures. Research this concept if necessary and then write an explanation of why the statement is erroneous. Include in your explanation the reason it would be correct to add 32 in certain cases.

2. Sanjeev considered nearly all of the problems in XinZin to be documentation management problems. However, the problems were finally resolved through cross-cultural training. What made Sanjeev believe proper documentation management would solve the problems? Was he really mistaken in believing in the need for better documentation management? Explain why or why not.

3. The situation at XinZin's Product Documentation Group (PDG) was obviously messy and complex. Do you think you can sort out any fundamental problem(s)? Is it personalities, language barriers, particular cultural differences, documentation practices, management and human resource practices, technical knowledge, power, or something else?

4. From Sanjeev to Xiaoxing, Peter, Tom, Meng, and Jao, all characters in the story seemed to have been both "the culprit and victim" of the problems at PDG. What do you make of these characters and their motives, concerns, and complex relationships? What false or correct moves did they make? Who, if anyone, should be the most responsible for preventing or solving the problems at PDG?

5. Sanjeev ended the story by commenting on "aliens" and "my people." What do you think of these two labels? Where do they come from and what do they signify? In what situations, for instance, has "aliens" been used to describe humans? And ultimately, what does this change of labels tell you about Sanjeev's transformation?

6. List the positive changes that occurred in PDG after the cross-cultural training was completed. Discuss the changes in terms of culture, language, technology, management, and so on. Which changes seem especially important? Why?

7. Discuss the potential outcomes and possible limitations of the kind of cross-cultural training program described in the story, whether at XinZin or other workplaces.

CHAPTER 8

CRASH OF CULTURES

Sam Racine

Sam J. Racine is the director of marketing for a For-tune 500 IT systems integrator. In this role, she spe-cializes in the design and facilitation of effective client conversations. Racine works with sales, account, and delivery teams to ensure they are fully prepared for client interactions and able to shape and positively in-fluence perceptions and collaborations. As part of this role, Racine manages the production of communication collaterals that accompany client conversations. These include the integration of new media and technology, from software to mobile devices to video use. Racine is an experienced facilitator, specialist in computer-mediated communication, and de-signer of human–computer interfaces. She is a Six Sigma Master Black Belt and is passionate about leveraging communication and change strategies to transform work environments. She also holds six patents, is a published researcher with a rec-ognized track record in professional journals and periodicals, and has been a fre-quent presenter/speaker at national and international conferences. Previous experi-ence includes participation in onsite service delivery for client accounts, client account planning, communication portals, and user interface design. Racine holds a Ph.D. in technical communication and human factors from the University of Min-nesota.

CHAPTER SYNOPSIS

She had never met the team she would have to bring together in a very short time in order to win a major contract with a large, private hospital in Austin, Texas. From the first hours they began to work together, Sam, who describes herself as a communication expert whose job is "deal support," began to question all she thought she knew about American culture. In her story she raises questions about the influence of geographic region, profession, gender, and age on cultural identity and how those factors came into conflict as the team members, who came from at least half a dozen different states and as many different fields, struggled to agree on winning themes (or as they say in the industry, "win themes"), team definitions, and how they would introduce themselves to the client. In the process she sees that for each of the 13 contractors she is trying to unite, some of those cultural factors may be more important than the others. Moreover, the influence of each factor may become more or less significant from moment to moment. It was a difficult task in team development for Sam but a tremendous learning experience in culture.

I'm a communication expert contracting for an IT firm on the east coast but living in the Midwest. I'm on the border between Wisconsin and Minnesota and so claim allegiance to both states, except during football season when my commitment stands rightly and justly for the green and gold (the Green Bay Packers for those of you who missed the 45th Super Bowl).

This story, however, happens in Austin, Texas. I tell you about Minnesota and Wisconsin not so you can marvel at the frequent-flyer points I accumulated the summer I worked there or grant me sympathy for leaving the lushness of the north in June to swelter in the intense humidity of Austin. No, I tell this story to show a lesson I learned, a lesson I didn't see coming but have taken with me to every job since. I learned that for some of us, the person we are, how we think of ourselves, and how we imagine our behavior are heavily influenced by where we live or from where we come. The regional affiliation, however, is not universal, nor is it always shared or respected. This is the story of how I came to learn about the importance of regionalism in consciously shaping behavior and how I came to use it and manage it, even as I questioned it and, yes, mocked it.

Before I go on, let me say something more about being from the Midwest: When we are from the Midwest, we do not volunteer this information when meeting people from the other parts of the country. Now I cannot say this is a rule for all, but it has been my study that for people from the Midwest, especially the northern Midwest, we only admit we are from the Midwest when explaining why 60 degree weather is not cold or why we have no sympathy for Washington DC, which closes down if snow accumulation reaches 2 inches. As Midwesterners we don't take credit for where we are from—at least not as much as people from some other parts of the country do.

MY WORK IN DEAL SUPPORT

I work for a firm that is a large IT services provider—we integrate computer systems and networks, provide both hardware and software, and do a fair amount of outsourcing for very large clients and governments worldwide. Our large clients are industry leaders: airlines, hospital networks, stock exchanges, and government services. In my role, I work all over the United States supporting client engagements, on projects where either my firm is onsite delivering a contract or we are trying to win a contract.

In this latter capacity, winning a contract, I'm assigned to what is called "deal support," which means I work with the proposal team to win the contract. The process begins when a client such as an industry organization releases a request for proposal (RFP for short) to a select group of vendors and/or to the public at large. If it is a proposal that interests my firm, we submit a written response to the RFP. The client scores the submissions according to its criteria and selects the highest scoring firms. This is called making "down select" and there are usually about three to five vendors selected and invited to come to the client's site. When the team from my company makes the down-select list and goes onsite, it is typically interviewed for a day. The client's organization members will critique our team while we provide demos, answer schedule and solution questions, discuss contract terms, and so forth. From the down-selected group of vendors, a contract winner is chosen. All this sounds simple enough, but remember these are multi-million-dollar, multiyear contracts so the RFP itself can be 200 pages in length, with our response typically at 600 pages. My company's writers for each proposal are usually a team of 12 or more, often the same group who will participate in the down-select interviews and manage delivery of the solution if they are chosen for the proposed work.

My role in this process often begins with the written response, where I help the team to discover and articulate the win themes to differentiate ourselves from competition. When we reach down select, the fun really starts: I travel to wherever the team is working, typically colocated with the client organization, and spend weeks working with the team to prepare for the down-select/interview process. I work with individuals to turn them into a cohesive team capable of excelling in the interviews and demonstrations, which are often face to face, live, and synchronous, usually with all parties physically in the same room. I draw attention to the live portion because look, readers, these are IT folks. Many of them are introverted, asocial, often loners, and I need to bring them together into a single, likeable team.

A TEXAN PROJECT

This story takes place during the summer of 2010 when I joined a team chasing a long-term, $100 million contract in Austin, Texas, for a private health agency. The RFP focused on the client's desire to have a systems integrator who can perform the IT work necessary to support the client's financial systems for its new research wing. Specifically, the organization wanted the vendor to integrate several existing financial systems into a new billing system. In addition to processing the billing, the system managed all the paperwork generated that creates an actual bill. It is astounding how much paperwork that is.

In my line of work chasing deals, context is everything, so I want to provide a little more background. In this trip to Texas, we were in down select for a health management agency. This was a private agency funded with public dollars that provides oversight and management for a group of hospitals throughout the state of Texas. As

I said, the client was looking for a systems integrator to provide a better system for processing their billing and documentation. "System," in this context, means better business rules and processing, better usability, less errors, and consistency with federal and state guidelines for privacy and data protection. And, of course, the client expected performance from the new system: no downtime, flexible systems, easy and inexpensive maintenance with protection against disaster. I was joining a team that had already submitted the written proposal. I had read the proposal, but I was not actively involved in generating its win themes—the value propositions that differentiate us from our competitors. In fact, the win themes were not necessarily articulated before I came along—we were in down select largely due to our technical expertise and, as some suspected, perhaps to keep a few of the other vendors competitive in their pricing.

The submitted proposal was primed by my firm—meaning we are the oversight firm, much like a general contractor—but included four subcontractors local to Texas, including historically underemployed businesses (HUBs) such as women- or Latina-led organizations. Although a private deal, the health agency wanted to follow public hiring guidelines for the state of Texas, hence the HUB participation. The funding for the solution included some state funds, so the private firm wanted to appear beyond compliance with state requirements. This means that my firm—a large integration firm—would require support from small businesses in order to meet the HUB requirements. Traditionally, systems integration staff are men, so I also expect to see women-owned and women-staffed firms brought into the picture to meet the state's requirements.

The written response was submitted one month prior. We were now down selected to one of three firms and the interview process would be a one-day, confidential discussion with the potential client. This down-select activity is very much like a job interview: The client was intending to evaluate our skills, our understanding of the solution, the solution itself, and our experience and credentials. These discussions were structured by the client to include introductions, a brief (15-minute) opening presentation by us, a longer presentation from the client, and then our answers to questions the client had provided. In down select, there is one key difference from a job interview: It is team based. The client organization needs to be impressed with the team as a whole, not just a single individual.

Unlike usual proposals which require some type of technical solution design in addition to delivery, in this case the technical solution was already determined. All three firms were bidding on the basis of doing the execution of the solution, not the solution itself. Thus, all three firms would be providing the same end state, using the same tools, and proceeding along the same time line. These parameters forced differentiation to be along two lines: price and suitability to deliver.

Price, always a tricky matter, became a nonissue for me to address. Significant work was already completed on determining the correct price to bid—making it palatable for the client and keeping my firm out of the red—but there was nothing in it that I had to either defend or counteract.

Suitability is another subject all together and essentially became the heart of my challenge: What is suitable to the Private Hospital of Austin, Department of Radiographic Research, to have the hospital confidently appreciate that we are the best team to handle its financial processing? Put plainly: What does the hospital want in this vendor? And once I know that, how do I make my team likeable?

We had only two weeks before we had to go in front of the client. Before my arrival, my prescribed role was to help the combined team of five total firms to sharpen our messages around win themes and ensure our answers were effective during the presentation. My tasks included helping the team construct and deliver the opening, coaching them on their introductions, preparing the written answers, and in general turning them into the one team the client wanted to employ. All of these required a foundational task of determining our win themes in order to differentiate ourselves from the two competitors. We needed to address potential or perceived weaknesses and in general construct and control the image of the team we wanted to convey.

A mere 45 minutes after my arrival, however, I had assumed the additional role of meeting facilitator, project manager, and timekeeper—this often happens to me on projects when we have subject matter specialists in technical fields but no communication experts. Soon enough, I also became the go-to person for lunch orders, printing requests, carpooling, fashion consultation, general arbitrator of disagreements, mother confessor, and inside informant. In short, I became the "cat herder" of our team for wrangling independent individuals into a forward-moving group.

MEETING TEAM MEMBERS

This was the first time that I would meet any of the team members, which included six fellow employees and eight subcontractors. It was also my first time in Austin, in the hotel, or in the conference site where we worked. One of my fellow employees I knew by reputation alone, but other than that they were newcomers to me. And, interestingly, it was the first time they had all met in person as well: Until this time, all meetings had been virtual. And only one of us was local to Austin. So, it was not just I who struggled to find a decent vegetarian restaurant that served cold beer; everyone was struggling with the logistics of life as much as with our presentation.

We met on a Tuesday morning, all of us positioned around the perimeter of a large conference room in the office of one of the subcontractors. There were fourteen of us: thirteen team members who would be in front of the client and myself, who was along for preparations but would not be in the room. (The client had put a limit on the number of individuals allowed to be present.) My fellow employees and I arrived 30 minutes before the meeting, already sweaty from the heat. On the drive into the office, we discovered something in common about us: Not one of us wanted to be in Texas. I'm not sure how we all came to agree to it, but it appeared all of us had been hurt by something or someone from Texas. From lost loves to lost football games to lost luggage, Texas had not been good to any of us. Roll into that the greetings from the hotel clerk, cab driver, and administrative assistant, who all noted that we "weren't from around here," we arrived at the meeting carrying a 100 percent humidity of grumpiness.

We were greeted by subcontractor staff, shown the bathrooms and stack of breakfast tortillas, and entered the conference room. I knew no one, of course, and watched as we settled in according to company affiliation. I moved across the room from my colleagues, listened to unproductive jabber for 45 minutes, and then took over. And by "took over" I mean I said to the room in general but with eye contact to the project manager, "So, how should we proceed to prepare? Do we have a plan for the next days? Do we have specific goals or a timeline?"

Without waiting for a response, I added "because if not, why don't we start there and I'm more than happy to facilitate that discussion. Would that work?" Of course as I was saying the last sentence, I was already rising from my seat walking to the whiteboard, so it made it easy for the group to agree it was a good idea because I was already in action. We worked out an agenda of what we needed to accomplish in two weeks, from introductions to win themes, from who takes the lead during question-and-answer sessions to providing written responses to questions the client provided. For what it is worth, these are no small tasks and in no time I realized I would be working the weekend.

The first step for me was deciding on the introductions our team would make to the client. Remember, my specific goal is to make these individuals into a cohesive, likable team, so I always take my time when we do our first introductions, encourage team members to interact, and take copious notes. I also needed to understand what our win themes would be and who could deliver them best, so I asked the team to introduce themselves, their role on the delivery project, and what they think the win theme should be. Here's what I learned:

Technical Architect, Colorado. Wicked smart, confident, a joker: "Elegance. An elegant solution delivered by someone other than Texans. Even Texans don't like Texans." I am serious. These are the first words from his mouth. I can't help myself but ask, in order to lighten the mood, why is it that he hates Texans so?

"Because they have no idea how to ski or how to drive and I'm sick of them ruining my slopes and my college football games."

I let the joke go but made a mental note to myself: Not sure this attitude will play well—client will likely want to keep the appearance of cordiality. I notice that as he makes this joke one side of the room deeply inhales but seems unwilling to exhale. Surely we all see this for the joke it is, right?

Project Manager, New York. A remarkably grumpy man whose cup was always half-empty, often literally: "What the client wants is experience and we don't have it so this whole thing is a waste. I'm here and I'll help, but we ain't gonna win."

Thank you, Mr. Ray of Sunshine, for your positive attitude. Note to myself: Must make him believe.

So far these are the first two voices in the room, and my now forced cheery demeanor does not feel well received.

Health Specialists, Indiana. Sincere, open faced and eager, joking about them being from "the heartland" of the country, and for the win themes say, "Sincerity. Commitment. Honesty."

I write down, "blah, blah."

There are at least two of them, maybe three. I'm not quite clear on the count because one of them distinguishes himself as being a "health systems" specialist and I'm unclear how that differs from a "health specialist" in general, but it seems to matter to him. I'm not surprised by their reference to "heartland." On the drive over, we had a lovely exchange based on the "don't mess with Texas" bumper stickers we saw on the streets. If such a bumper sticker is made for Indiana, they say, it would be something like "Leave our hearts alone."

Research Specialist, Philadelphia. The self-proclaimed "old man" of the group and a former hospital employee, engaged by my firm to bring insider information and ease transition to new providers: "Experience, longevity, knowledge. Been there, done it, will do it again and again."

Granted. I also note that he has described the win theme as being the sole reason he had been engaged in this project. I'm not so blind I can't see this as self-serving: If we focus on longevity, he is of value and his paycheck goes up.

Rock Star Technologists, Texas. Three young guns, hot shots, born and bred in Texas (their words): "Being stoked man, totally stoked and jazzed."

OK. I contemplate whether I would ever admit I was born and bred in any state. I'm starting to sense there is something I am missing here. Would you claim where you were born and bred? And if so, would that affiliation somehow set you apart from someone else? Is this simply a challenge to the Colorado technical architect (and his joke), since by title the architect outranks them. Or is it a Colorado-versus-Texas competition?

Captain of the Ship, Austin, Texas. Unchallenged leader, account manager, and the person who will receive commission. Imposing, with killer ability to shut down conversations with his cold hard glance. The one member of my company's team who is local.

He says nothing but gives a look that says he is evaluating these characters as fast as I am. He also says on several occasions related to fashion choices, lunch choices, and code of conduct that "this is Texas." I have yet to translate that into a meaningful set of directions but, again, sense there is something here I better come to understand and articulate and do so quickly. Does Texas have a code of conduct for dressing beyond boots and belt buckles? And for food choices beyond barbeque? I thought barbeque was owned by Georgia and Kansas.

Family Support Specialists, Texas. Two women, from a woman-owned business. Except for people I have not yet met who may yet appear at the interview, these are the only two women in the room so far. They are Texas based but don't strike me as being life-long residents. From the start I struggle to hear the differences in their voices and can't quite determine who is who. When either one of them speaks she turns to the other, who is sitting right next to her, nodding support and making direct eye contact, once placing her hand on the other's forearm for support. I like them instantly as I would a pair of aunts who bring my favorite pies for family events. After introducing themselves and the mission of the corporation, which is one of active engagement and respect for all voices, they pronounce the secret to being likeable as "collaboration."

I can't help myself and while I hate myself for doing it, I roll my eyes and make a notation: Of course. And, of course, these collaborators just have to be women, don't they? I'm not sure if I would have been less disappointed if at least they had launched into a Texas-is-all-that-and-more tirade instead of this so very expected collaborative stance. I notice that as they start to speak the other attendees in the room, all men, start to check their e-mail. I don't blame them—there is something so token about including these two women that it feels insulting.

Programmer, Texas. Back row. In fact, in a corner behind the back row, facing away from the team. I have to call on him to get his attention. A Texan, for whom the descriptors "tall drink of water" and "still waters run deep" were apparently crafted since he's not so much folded into a slump but rather a fluid line of back-and-forth folds that leaves his head, despite his six-foot plus frame, only inches above his keyboard. The tallest man in the room ends up only two inches taller than his computer so we can't actually see his eyes. His contribution is to announce: "Texans buy from Texans. Or from Mexico, because that's the same thing."

Note to self: Really? There's a pattern forming here. I expected that we would have personnel from Texas. We always try to have a local crew to save travel costs and public-funded projects want local staff. But there is something else in this room that sounds like pride and I can't discover the source. And the programmer is echoing what the captain has already put forth.

And so I have my team for the down select: the architect who will design the technical solution, the researcher who will ensure the solution aligns with the legal and regulatory requirements including privacy and protection of sensitive data, the technologists who will implement the solution, the health and family support specialists who will design the business process using their experience with the domain, and of course the project manager, who will keep it all together, the captain, who will interact with the client's executives, and the programmer, who will provide all the code. These are the folks who will be in front of the client and manage the main delivery should we win the contract and will also oversee the 40+ individuals who will work on this project once we win it.

As we finish the introductions, the room opens up to a bit of chatter—"you're taller than I expected" ... "how's that dog of yours doing" ... "you're from the north, right?" All phrases that point to team members who have only interacted via phone. I also hear comments about Austin as the "blue island in the sea of red,"* recommendations for dinner, a few remarks about professional certifications and education, and a little bit of posturing and positioning one's credentials.

We then break up into small teams to discuss the solution. During this time I take a breather and start to sketch the dynamics of the team as it is and how I need it to appear for the client. I try to understand each individual's self-identification: those who tell me they are from Texas, those who only mention their home to demonstrate they are not from Texas, and those who are puzzled why we mention states at all. Some clearly consider their profession to be their defining characteristic, or their years of experience, or their political affiliation, or their gender. I wonder what it is that makes me who I am. Or, better said, I wonder what I can credit as making myself who I am.

DETERMINE WIN THEMES

After the small group discussion, the team comes back together to work out the win themes. I need these determined first in order to build the rest of our responses

---

*For readers not familiar with American politics, a city/state is considered blue if the voting history favors the Democratic party and red if it favors the Republican party.

around them and to ensure we can support the claims we make together. I first ask the team to tell me what we want to say about ourselves as a team that will matter. The group starts throwing out phrases with the rock stars challenging the architect, who is quite capable of holding his own. I sense a little competition has already begun:

Rock stars: "One throat to choke."

Architect: "Lift and shift baby."

Rock stars: "Keeping it weird."

We have already identified this during the drive in as an Austin-based anthem, and I know the architect views it with contempt. He levels his eyes at the rock stars as if he's throwing down a winning hand of cards and says, "Won and done."

Not to be outdone, the rock stars say, "Texas based."

The captain and the programmer agree, as does the researcher, except he corrects the statement to be "Austin based."

"Working together." "Communicating together." These last two are volunteered by the two-person team I now call the collaborators. I note, not for the last time, that their comments seem only directed to each other.

And, somehow, we go from analyzing the client to:

"Are we having a group picture taken?"

"Are we wearing suits?"

"Do the women wear skirts or pantsuits?"

"Do we have enough women?"

"I don't know about you but I've had enough women..."

This is a signal that we're starting to digress. However, we all have to get to know each other so I let the exchange go a little longer, until the room starts to feel unsettled. I notice the pleading glances from the project manager, the slurping of coffee from the dregs of the mug, and the reemergence of several laptop screens. I say, "Ok ... let's sort this out."

At this moment, the captain stands and moves to the center of the room, puts on his glasses, turns to the team, and takes the glasses off again: "We're here because we care about the research this hospital is conducting. It can save lives, especially the lives of young children struggling with terrible illnesses. Children are our most valued gifts and we need to protect them. In Texas we believe in such things...." The speech goes on but the ending is remarkable, at least to us non-Texans: "God Bless America and God Bless Texas," at which point the architect bursts into hysterical laughter claiming, "I almost believed you until you blessed the state."

Meanwhile, two of the rock stars yell "right on" and the Midwesterns look overwhelmed by this display of emotion. The collaborators look like they're going to tear up and squeeze each others' hands. The third rock star has put his hand over his heart at this blessing and leaves it there, to which the New York project manager notes, "Are you having a heart attack or something?" I scan the room: Half are moved while the others are in various poses of disbelief and mockery, and we are less than 12 minutes into the deciding-win-themes exercise.

After battling my own confusion for a few minutes longer, I'm forced to say, "Is it a big deal to be from Texas?" This question is met with a series of hoots, some commentary by the architect on a Texan's ability to ski, and a look of grave concern from every local Texan. I do not comprehend and ask again.

"Honey," one of the collaborators, "being from Texas is all that matters." I am stunned, there are more sputters and now hesitant laughter, and I feel that somehow a gauntlet has been thrown down.

The captain speaks, "There are two things we don't joke about in Texas: God and Texas."

Oh. Am I in a Sandra Bullock movie? I can only imagine the conversation that will happen in the ride back to the hotel among the non-Texans.

Providing a local team does matter—public funds are being used and states want them to go to their own personnel. I understand that, but there seems to be something much deeper here. I think to myself, not quite in horror but with deep concern, not only does half of the team believe one thing while the other half believes something else but neither side respects the other. I struggle to manage the next exchange, trying to keep the honest emotion that bonds half the room while not excluding the other, when the programmer states, directly and clearly, "I've had enough women in my life." I now know he'll discuss Texas and women and in that order. The Texas theme continues to surface and I see that it is not simply a running gag to make us non-Texans comfortable. We are in Texas and Texans care about being in Texas. I'm from Minnesota, does anyone care?

We must come up with a comprehensive win theme platform so I walk to the front of the room, take a pen to the whiteboard, and write: Here's what matters— here are the win themes you have thrown out so far this morning:

- Elegance, technically savvy
- Local, but not local (?)
- Experience, which we don't have
- Youth (?), in touch with the times
- Emotional play—sincerity, honesty, apple pie
- Collaboration, conduct
- Team unity
- True (?) commitment to kids (demonstrated?)

I ask where they want to begin, get a series of blank looks, so I kick it off myself: "Local but not local, where do we stand on this?"

The captain sits back, removes his glasses, looks momentarily at each person, at the architect, and states: "Texans buy from Texans. And we're here in Austin, not in Dallas, so don't make that mistake again. Let's get that straight and move on."

The programmer nods his head and continues to look out the window. I cannot simply leave it here—half the room does not understand what Dallas has to do with anything so cannot be committed to the theme. Without commitment, it is only words. And these are not actors: If they do not believe, they cannot deliver.

I ask for clarification: "What part do we need to get straight? Texas or Austin?"

Captain: "In Texas, we take care of our own. We do everything in our power to support each other and to keep our money here or in Mexico. We have a history of being robbed financially and so when there are state funds on the table they must go local or it's all over the papers and then the governor gets angry and the administration changes when the next election comes around. Texans must buy from Texans. Or we've got problems. And, besides, it's the right thing to do, and it's how we do business here."

OK, I can buy that. I see the health specialists nod and the project manager shrug a "makes sense," but the architect isn't giving in.

"Isn't Dallas the same as Austin?" I ask.

I am met with a series of groans. Dallas, apparently, is the land of "transplants," "pseudocowboys," and "cosmopolitan urbanites." Austin is the land of independence, the state capital, the place where power resides. The rock stars say to me slyly, "that's why we keep it weird here." Oh, I get it. It's the independence, the fierce independence, upon which they brand themselves.

I check off this win theme and write "Local, Austin-based staff." So we want to portray a local staff when part of a solution will be funded with state funds. This is not unusual. All states want to support their local citizenry. Still, there is something more happening here, a regional identification with the state that I have not experienced anywhere else. Do Wisconsinites believe in themselves as Wisconsinites? Minnesotans? Illinoisans? No. The Texas brand is better defined, better shared, and more desirable. What is the brand for Wisconsin—independence, ruggedness, support of neighbors? No, we wear cheesehead hats. Maybe Texas is on to something.

"Youth? Energy? Where are we on this one?" I continue.

The researcher, from his corner seat in the back right, calmly and excruciatingly slowly says, "I've been serving this client for 35 years. If you want someone younger, then I'm leaving now." I look around the room. No one makes eye contact.

"Well," I say, "let's unpack this term of youth. What do we mean by it?" Relieved of the tension, answers fly:

"Energy."

"Vitality."

"Freshness." I groan inwardly at what sounds like a menu for a salad bar.

"Not tied to old ways."

"Not tied to expensive ways." I suspect we are getting somewhere.

"Not tied to expensive women."

And with that, the running gag is invented. I let it run. Apparently it is more fun to think of women as being expensive than as being, say, smart. Who am I to change this world?

In the end, we decide what we are bringing here is creativity, which I force them to translate into something real and actionable and we decide that the tracking of creativity and provision of brainstorming sessions to generate new project ideas would be a regular topic on the project management office's agenda.

We continue to pick the win themes off, one by one. The hours pass and we define (easily) collaboration and the importance of appearing as one team and not five. Our two family health specialists do the lion's share of speaking here and seem unconcerned that no one else contributes. The rest of the team seems content to check e-mail—again—while the ladies happily chatter on. Eventually I ban the use of the negative "one throat to choke" and instead we move to "one team" and assign the Midwesterns the side activity of making the phrase sound more sincere and honest. They return after lunch with the modification from "one team" to "your team." Simple but elegant: The architect likes it.

We then tackle experience. On a giant white board, we form a detailed list of each company's and each individual's related experience. We elaborate. We map connections. We start to see some depth, some overlap, and what starts to feel like a strong picture. I look at the clock—it is just before our end-of-day break—and decide it is time to force an outstanding issue. After evaluating the room of people and the emotional play we need in order to have the client like us, I make an intentional move to introduce two crucial concepts into our mix: "belief" and "us." I need to turn them from a collection of 13 individuals into one team, and I need to start it today, the first day, before we learn to interact as individuals only. If that happens, I'll have to undo the work in order to form a team. I know it is the moment to introduce believing in ourselves just as I know when a storm is coming. It is not genius but the recognition of a series of small indicators that, put together, lead to only one conclusion: It is the time to turn them into a collective whole. If we continue thinking of ourselves as individuals much longer we will not become a team. We have spent enough time talking about our individual capabilities to earn each others' respect, to position ourselves against each other, to indicate our likes and irritations.

I look at the project manager and say, "Do you believe we can win? Because if you don't believe we can, there's not enough coaching in the world to turn us into a strong performance next week. You're not actors."

The move hangs on the project manager and he is smart enough to know that, regardless of whether he believes or not, we can't move forward until he commits. It is a trap of course, but he walks into it because he is smart and he has been in this situation before. He looks at me, looks at the team, and with a little too much mumbling for my taste, says, "Yeah, we can do it. Yeah." Ok, it is not a Rocky Balboa moment but I take it. I push him to articulate that he believes this team will succeed and so I make him repeat it:

"Do you believe we can do it?"

"Yes, I do believe we can."

In closing, he adds, "But with all due respect to the few women we do have here, the thing I most believe is that, yes, we should be tied to women, but not expensive women." The room guffaws and breaks for the last respite until the end of the day.

As we trudge back to our seats in the conference room, the programmer hands me a note stating: "True commitment doesn't matter if it's not demonstrated." I thank him, contemplate some joke about how many programmers does it take to discuss changing a light bulb (one or one hundred, no difference in the amount of conversation), and plot out the day's last activity to create a positioning statement.

We finish the day with a rough statement of our team's definition: "We are your team: experienced, local, and committed. We are also creative and looking forward to working with you, co-managing challenges, and bringing new solutions." I am getting nods and some fatigued looks. When pressed to see if we have it all covered, the architect nearly shouts that we need to be "elegant."

OK. I suggest we unpack that word and take it to something meaningful and he says no: "Every architect in the room knows this word. We don't have to elaborate."

It is not lost on me that he is the only architect in the room, and it is also not lost on me that the rock star technologists, who will have to implement his designs, do not make eye contact. This hierarchy—common in a technical solution—is clearly in play and confirmed here: Architect is a superior role with superior expertise. What he states should be understood by all, but I wonder if everyone does understand, including the client. Since the rock stars are silenced, I search the room for input and the only eye contact is from the programmer, who replies, "This is true." Fearing another note, I accept his input and "elegant" stays.

Before breaking for the day, we assign our next task of writing our introductory statements. We determine the overall agenda and length of the statement. Some team members feel it is important to take enough time to make an impact. Others believe it is best to spit out their name and sit down, letting "their résumés talk." (I want to screech at this point, "your résumés do not have voices.") In the end we agree they will state their names, their roles, why they are excited about being part of this team, and any connection they have to Texas, given that we have accepted that the local-based play matters. I insist on having them write out their answers and we adjourn.

On the way to the hotel, the nonlocals share a single rental car to keep costs down. We make a game out of identifying how many lone stars we can find outside the windows. To our somewhat horror, they are everywhere: embossed in cement on all highway structures, adorning bumpers and gas tanks, on hundreds of restaurants and service agencies, on homes with a blue tarp for a roof but a slate star next to the mailbox. I ask my fellow travelers if they can name the slogan for their state or even describe the state flag. We are pretty sure California has a bear on its flag, but we are not 100 percent sure. We cannot even recall the background for Minnesota's. We don't wear our statehoods like a badge of honor as the Texans do. We are confused by their identity and slightly intimidated. A voice from the back seat asks, "are we missing something?" and I wonder if he means we're missing the allure or missing the community and shared identification that the Texans have. Are we jealous?

PRACTICE RUN

We reconvene in the morning and map out our schedule for the day, which includes time to practice our introductions. We carefully look at the client's agenda, the time allotted to us, and make decisions about how much time each of us can have. This is not a simple math problem, as one might imagine, because—although I don't like to put it this way—some people in the room are more important than others and therefore require more time. The principal speakers—the captain and the president of the firm with the two family specialists—one of whom happens to be a woman, get 2 minutes and the rest of the staff receives 90 seconds.

We schedule the practice run for after lunch and I ask who has written their introduction verbatim. I receive assents from half in the room and put on my teacherly voice and take a minute to explain that we need the scripts written so we can carefully edit them. When the research specialist makes it clear that a man of his age does not need to write out his introduction since he has been doing it for years, I use an old technique to make him comply: I beg. There is some moaning and turning away but I do not leave the front of the room until I have agreement. Since I am not here to make friends, I am not bothered by a lack of warm reception.

During our practice run, I am confronted by yet another difference which I choose to support rather than quell. I allow everyone to use the medium that gives them comfort: handwritten for some, napkin backs for others (truly the rock stars just can't help themselves), laptops, and smart phones. Again, I am struck by how stereotypes ring true: Technical people use the most technically advance media— iPhones and BlackBerry® smartphones. Technologists disdain popular techniques and rely on coolness as a way to distinguish themselves from "the others." Programmers can speak to their computers but not to another person. Do different professions attract different personalities or do personalities develop once in the professions? I think of my own profession: As communicators, are there things we must always do, besides correct people's English and note that we have "read the book" instead of watched the movie?

I ask each to read their introduction aloud, I time them, and I allow each of them to proceed in their own way. Some edit on the fly, some are rigidly rehearsed, some stand at attention, and some are unable to face the crowd. At the end of each presentation, I ask the team to critique. All are overly long, a few are on the mark in both text and presentation, but most are woefully inadequate. I make a list of words I'm hearing.

At the end of the performance, I ask the team to discuss and make suggestions for team improvement. Cautiously and slowly, we come together in agreements:

"Let's make sure to use the term 'your' and 'you.' And talk directly to them."

"Let's stand when we speak."

"We need to pause between our first and surnames."

I draw attention to the list of greetings I have noticed: "hello," "hey," "good to meet you," "nice to see you all," and "nice to see y'all." And I make a list of their appeals: Personnel are 'thrilled," "excited," and "pleased." Members are being "totally stoked," "incredibly geeked," and "ready to take it on."

We focus on the technology expert who is ready to rock their world and uses a fist pump in his closing. I ask the group, does this work? We discuss formality, expectations for what our competitors will provide, and standard conventions for closing. What emerges is that the technology expert should tone it down. He looks at us and does a robotic routine in which he moves with choppy arm movements and states, in flat, uninflected, and mechanical words, "My name is Robot. I am an automaton. I am not allowed to express my individuality." We laugh, and I remember a very old argument of a melting pot versus individuality. Have we gone too far? Does being a team mean being all the same? Where does the line of expression bend to meet the line of convention? I find myself remarkably unsettled, and while we laugh at the technologist's impression, he is clearly hurt. He agrees to our input. We take a breather and move on to other messaging.

The struggles around positioning continue over the next days. We agree to wear suit coats and ties, but do the shirts have to be white? The project manager prefers light blue because his spouse has told him it warms his face. Can the women attendees wear a patterned suit or "are we to appear as little men in pinstripes and straight cuts?" Do we introduce ourselves by our functional areas or by how we are seated in the room? Do we mention the companies from which we come or do we refer to branding ourselves as one team? How do we position old guys relative to young guys, women to men, and locals to those from a distance?

Through it all, I agonize that if we cut too much and force too much, our team members will lack confidence in who they are, will be unable to express true commitment and concern, and in the end *will* appear as robots. To be seen as a team we need to be similar, but should we appear so similar that we cease to be individuals?

In the end, we decide that the principal speakers will introduce their respective firms. This will be followed by the statement: "That's the last you'll hear of our individual affiliations because from this point forward, we only exist as one team." The captain will close with the blessing of Texas and the architect and project manager will not flinch or allow their eyelids to flutter. The technologist can close with "We're totally psyched (but not stoked)," provided there is general good cheer and handshaking when they leave. The collaborators are still not collaborating but merely talking to each other and no one else, but I keep my mouth shut because they are terribly warm and sweet.

After reflecting on our experience, the captain makes the decision to include as part of his opening remarks how we came as individuals with differences—sometimes subtle and sometimes openly antagonistic—but we worked and bonded and became the team the client sees, a group melding individual experiences into a cohesive whole.

I reflect that we are who we are and what defines us is different for each of us. For some, that regional affiliation is everything. Why? I do not know. Is it because of the brand itself or because we want to be from somewhere? For others, it is our expertise: I suspect that even if the architect had been born and bred in Texas, what matters to him is his professional standing and not the state in which he lives. And for others, it is their way of being—like the collaborators—or experience in life— like the researcher. And of course it is never just one thing that shapes any of us. As for me, I still don't know the Minnesota motto. (Does it have one?)

DAY OF PRESENTATION

The morning of the presentation we convene in a hotel across from the client site. On the way there the architect has found a record-breaking 12 lone star graphics. There is a moment when the rock stars feel mocked, a touch of tension, but they let it go as one of them discusses the new mosaic star he is planning to add to his back garden. The project manager offers to help once we win the deal. But he "ain't blessing anything unless there's alcohol there." We all laugh, we take a breath, and we do the final inspection of teeth, ties, garment openings, and hosiery. I hear the programmer practicing his own name and I reassure him he has the pace perfect. We line up as planned. I wish them well as they move behind client security and say a small prayer to God, if she's not too busy blessing Texas, to please bless us.

I do not know how it went from there. I wasn't allowed to participate or to monitor. I cannot tell you who missed their lines, whether the client bought into our win themes, or whether any of us outsiders smirked at blessing Texas. What I can tell you is that night I received a call from what I can only describe as a team, because there were so many voices at once I could not discern one from another. They were indeed "stoked," and they were happy and pleased. In their minds, "we hit it." And I guess they did because we were invited back to resubmit our initial proposal based on the interview process. Only this time, we were down selected to only two. And the winner is....

I can also tell you that what I learned about the importance of regional culture on this project is a factor I continue to consider in all of my work facilitating team development. Whether it is Midwestern or Texan, regional culture helps to shape who we are. This is not to say that all people from the same region would think or act the same way, because various factors, including profession, gender, life experience, and other personal factors, also play important roles. But it does mean cultural consideration is important even in facilitating "domestic" projects where all team members apparently come from the same nation state.

RECOMMENDED READINGS

1. For more information on working in and communicating with cross-functional and cross-cultural teams, refer to the following sources:

   - R. Kent-Drury, "Bridging boundaries, negotiating differences: The nature of leadership in cross-functional proposal-writing groups," *IEEE Transactions on Professional Communication,* vol. 43, no. 1, pp. 90–98, 2000.

   - N. P. Napier and R. D, Johnson, "Technical projects: Understanding teamwork satisfaction in an introductory IS course," *Journal of Information Systems Education,* vol. 18, no. 1, pp. 39–48, 2007.

   - R. T. Barker and L. Zifcak, "Communication and gender in workplace 2000: Creating a contextually-based integrated paradigm," *Journal of Technical Writing and Communication,* vol. 29, no. 4, pp. 335–347, 1999.

   - L. McGee, "Communication channels used by technical writers throughout the documentation process," *IEEE Transactions on Professional Communication,* vol. 43, no. 1, pp. 35–50, February 2000.

   - R. Rosenfeld and J. Euchner, "Culture, people, and innovation: An interview with Robert Rosenfeld," *Research Technology Management,* vol. 55, no. 2, pp. 13–17, March–April, 2012.

2. For more information on the U.S. Midwestern and Texan cultures, refer to the following sources:

   - R. Sisson, C. Zacher, and A. Cayton, Eds., *The American Midwest: An Interpretive Encyclopedia.* Bloomington, IN: Indiana University Press, 2006.
   - J. R. Shortridg, *The Middle West: Its Meaning in American Culture.* Lawrence, KS: University Press of Kansas, 1989.
   - M. Ivins, *Molly Ivins Can't Say That, Can She?* New York: Vintage Books, 1992.
   - L. Clemons, *Branding Texas: Performing Culture in the Lone Star State.* Austin, TX: University of Texas Press, 2008.

DISCUSSION QUESTIONS

1. Discuss your own experience working with people from other regions of your native country or people with different professional backgrounds from yours. What difficulties have you encountered in such interactions? Do you find there are assumptions you can rely on for being able to understand or work with people from certain regions of your country, of certain ages, of certain genders or sexual identities, in certain professions, or in certain social or organizational roles? Discuss those assumptions.

2. Sam identifies the "cast of characters" in her story by their professional roles. Keep in mind that she uses humor throughout the story. How do these character labels function in the story? Are they simply convenient ways to help readers visualize and keep separate all of the many characters in the story? Are they a type of stereotype? If they are stereotypes, do you think she is aware they could be perceived as unfair oversimplifications?

3. The programmer from Texas makes the comment, "Texans buy from Texans. Or from Mexico, because that's the same thing." Sam expresses incredulity at this, and to many readers it is undoubtedly a strange thing to say. What assumptions do you think the programmer is making? Is he expressing a political view? An economic reality? A sense of cultural unity between Texas and Mexico?

4. Consider those stereotypes of people (who are different from you) that are commonly held in your social/cultural circle. Do these perspectives influence the way you interact with or perceive those people? Have you ever had to make a conscious effort not to stereotype someone? Is it really possible to see or understand everyone you meet as a unique individual to whom no generalizations, assumptions, or stereotypes apply?

5. Does cultural identity, whether regional or professional, become less important for the team members as they carry out their task? Do the team members'

cultural identities change through the team-building process? Does the team develop a cultural identity of its own? Discuss and explain your answers.

6. Reflect on and discuss your own cultural identity or identities. How influential are those identities in different situations?

7. Which characters in the story do you tend to identify with most? Do you identify with those characters most because of their regional culture, gender, or professional identity or for other reasons? Explain.

8. Try using Geert Hofstede's cultural dimensions to analyze the cultural information about each character in this story. You can find Hofstede's model on websites such as http://www.clearlycultural.com/geert-hofstede-cultural-dimensions/. How well does the model work in explaining the behaviors of the characters?

CHAPTER 9

COLLABORATION ON A PAN-EUROPEAN PROJECT SPANNING 35 NATIONAL RESEARCH AND EDUCATION NETWORKS ACROSS THE EUROPEAN UNION

Warren Singer

Warren Singer is a regular contributor of articles to Communicator, *the quarterly magazine of the Institute of Scientific and Technical Communicators (ISTC) and has contributed to* Intercom, *the monthly magazine of the Society for Technical Communication (STC). Singer has over 16 years' experience as a technical communicator, much of this spent working in multicultural environments for international organizations and businesses in South Africa, New Zealand, Israel, and the United Kingdom. He is a founding partner of Cambridge Technical Communicators, a technical communications company based in Cambridge.*

CHAPTER SYNOPSIS

Hired to help write a high-stakes annual report for GÉANT, freelance technical writer Warren Singer quickly found that he needed to collaborate with many people representing multiple organizations, roles, languages, and cultures. GÉANT is a data network linking scientific, technical, and cultural research organizations across Europe for users over much of the world. Warren's collaborators included clerical staff, other technical writers, senior directors, project managers, and engineers. Many of these people represented different participating organizations in the GÉANT network, but their participation did not necessarily entail working within a strong administrative structure. For this reason, their collaboration did not mean they were obliged to cooperate. And although everyone spoke English, they were not all fluent speakers of English. Collaborators often communicated and worked together by conference calls, only occasionally meeting face to face. Differences based on national origin, organizational structure and culture, international political histories, professional differences, age, and personalities and interpersonal dynamics influenced the ways participants interacted—or sometimes conflicted. Inevitably, different ways of understanding the task, different investments in the task, and the growing pressure of deadlines complicated what might seem to have been a routine technical writing project despite its intrinsic complexity and high stakes.

Warren's story reveals the multiple skills he needed to establish trust and understanding, to mediate conflicts among other team members, to negotiate meanings and purposes in the developing text, and to continually check his own assumptions about the ways culture and national origin may relate to and differ for individual character and personality.

JOINING ADVANCED NETWORK TECHNOLOGY (ANT)

First day on the project. It was a bright, sunny day in April 2010. Driving up the parking barrier at the ANT location (all company and character names used are pseudonyms) on Hills Road, Cambridge, I could see the spring flowers were out and blooming. After asking a security guard, I parked my car in the allocated parking space, visitor parking bay 4, and walked to the entrance of ANT on the third floor. The inside of the building was modern and clean, with impressive statues put out in the hall and at the top of the spiral staircase.

I had been living in Cambridge, the United Kingdom, for about 10 years, working primarily as a freelancer on technical communications projects, before joining ANT. Responding to an advert posted on Jobserve, a popular website for advertising IT positions, I was hired as a technical author on a short-term (six-month) contract.

ANT was the team responsible for project managing Gigabit European Academic Network Technology (GÉANT), a gigantic 40-gigabit fibre-optic network dedicated to research and education networks across Europe, with global connections to similar networks in Canada, the United States, Africa, South America, Asia and the Middle East. GÉANT has had an impact on many of the major research and advanced technological projects in Europe and is used by world-renowned scientific research projects, such as the Large Hadron Collider in Conseil Européen pour la Recherche Nucléaire (CERN) and the Distributed European Infrastructure for Supercomputing Applications (DEISA). The GÉANT project was a truly multinational endeavour. The teams that made up the different member organisations came from across Europe (If you want to find out more about this fascinating, pan-European project, take a look at the GÉANT website at http://www.geant.net).

ANT had an important role in planning the development of the GÉANT network and liaising with the 35 European National Research and Education Networks (NRENs) and partners as well as producing a stream of reports each year. Although it was a small unit with less than 50 employees, ANT likewise had a diverse workforce made up of employees from Spain, Poland, Italy, France, Germany, the Netherlands, the United Kingdom, Serbia and many more countries.

Ivan, the project coordinator, welcomed me warmly. Ivan was a tall, dark-haired Serbian who came across as being very calm, collected and confident. I had managed to impress him during the job interview. Ivan sat down with me and explained how activities within the GÉANT project were organised.

"The project is organised into a number of key areas or activities. Each activity is led by an activity leader. The activity is broken down into tasks, and each task has a task leader," Ivan explained.

"How many activities are there?" I asked.

"There are about sixteen," Ivan replied. "The leaders are taken from the project's NREN partners."

My understanding of the project was limited. I had read on the website that GÉANT was cofunded by the NRENs and the European Commission (EC) and was part of the EC's Research and Development (R&D) programme. The leaders from these organisations, who had established and now ran GÉANT, came from the various nationalities and cultural backgrounds across Europe. I wondered if language and culture would be an issue in working on this project.

"Are there any problems with language and culture that I should be aware of?" I asked.

Ivan sat back, thinking for a moment, then explained: "The activity leaders are taken from across Europe—some of them work for ANT, but many work for their own national organisations. We have people from Spain, Italy, Germany, the Netherlands, Sweden, France, Ireland, Bosnia, Serbia, the United Kingdom—you name it. Most speak English quite well, but not all. We have one or two that really struggle with English. We have to be quite diplomatic when talking to the activity leaders that work for other organisations, as they report to their own organisation and don't report directly to us."

I nodded. "So, what do the technical authors do?"

"We have a schedule of reports that have to be delivered to the EC each month," Ivan explained. "It's quite a heavy workload for the technical authors. They help the activity leaders prepare the reports and then we submit them to the EC for publication."

"Do the authors write the reports from scratch?" I asked.

"Not exactly," Ivan replied. "Although sometimes they may. Usually, they'll work with the activity leader to get a draft out, or the activity leader will send a draft for them to edit. Obviously, the owner of the report is the activity leader—they are responsible for reviewing and authorising the report for release."

"Is it just the reports that the authors work on?" I asked.

"No. The authors help with business cases, technical guides, marketing material and the websites. We'll get you involved in those as well."

"But one primary task for you, as we have talked about in the interview," Ivan added, "is to put together the annual report for the third generation of the GÉANT network project."

Yes, I had heard about this "annual report" quite a few times by now. It is supposed to be a 160+ page report, written by contributors within the 35 NRENs from across Europe. With a project budget of 250 million euros, the stakes were high. Managing the expectations of key stakeholders, obtaining reviews and producing within the space of a few weeks was a daunting task—compounded by the fact that contributors were dispersed across many different countries in Europe, spoke different languages and worked for different organisations.

Ivan stood up and smiled broadly. "Now let's get you introduced to the rest of the team."

He took me round to the cubicle where I would spend the next few months. Jenny, one of the technical authors, was originally from Germany but had settled in the United Kingdom. She was listening to music on her earphones and looked up as I was introduced. Jenny smiled warmly and shook my hand. She was dressed casually, which was typical in our work environment, wearing blue jeans and brown boots, but looked smart and well-groomed.

Ivan next introduced me to Susan, the other technical author, an English lady. Susan smiled briefly, shook my hand, then went back to her work. Paul, a training manager who ran courses on GÉANT services and products, also sat in our area and shook hands with me warmly.

After I had settled into my desk, Jenny and Susan took me to one of the meeting rooms and explained to me where all the files were located on the network and the templates I would need to get started.

During that first week, Ivan had me involved in several urgent deliverables: a technical paper describing the multi-domain management service for the Large Hadron Collider at CERN, a report discussing the progress and meetings with European scientific institutes and user groups wishing to join GÉANT, a business case for a new service and the technical annex detailing the project deliverables for the next year. Starting on these early documents gave me an opportunity to learn more about the project, get to know some of my team members and become comfortable with the document templates and processes put in place. All these would help me later on when preparing that annual report.

During these early days, I also tried to engage with the other technical authors in the team but found it difficult going. Jenny, in particular, seemed to me of a somewhat nervous disposition. I couldn't quite put my finger on it. The atmosphere in the office was relaxed, but Jenny seemed to be tremendously busy all the time. I suspected this was partly due to her personality. Having to travel a long distance each morning and then leaving late may have contributed to her somewhat impatient disposition. Paul seemed to be the most approachable of the three team members who sat in my area.

And on came Friday. I was about to survive my first week on the job. Standing in the kitchen, I was waiting for the kettle to boil to make tea and talking to Paul. It was lunchtime, and someone was heating up a curry in the microwave. The strong, spicy smell made my mouth water.

"What's up with Jenny?" I asked. "She's always busy with her drafts and the other day I tried to talk to her, and she just cut me off and rushed off in the middle to speak to someone else."

"Give it time," Paul suggested, "Jenny is very busy at the moment."

WEEK 2: WRITING A REPORT FOR CERN

"So tell me, Stefan, what's this report really about?"

Stefan was a somewhat shy and reserved young engineer from Bosnia and Herzegovina who had been working for ANT for several years. His role was managing the testing activities on the GÉANT network as well as deploying the trial network monitoring application his report was describing. I sat with Stefan in his little cubicle, with a draft of his report in my hand. I had read through the report and marked up comments, so I already had some understanding, but I wanted to hear more about it in his own words.

"It is complicated, but I try," Stefan began. He talked for a while in halting English, struggling for words. I listened carefully, then looked at the notes I had made in the margin of his report.

On the first page he had written: The MDM comprises of two appliances, for network measurements and passive measurements, for monitoring network delay and regular passive measurements for interface utilisations, input errors and discarded packet statistics network elements.

I had scribbled the following note next to this: The multidomain service has two types of "applications" or modules that perform monitoring of the network. Clarify with Stefan the role and purpose of each.

This was one of my first tasks— to edit a technical paper on a collaborative project with CERN, which was running the Large Hadron Collider in Switzerland. The hadron collider is a two-mile-long, underground, circular particle accelerator which accelerates atoms to near the speed of light before smashing them together in an attempt to emulate conditions that may have existed during the big bang. The collider became famous in 2012 for the discovery of the Higgs-boson particle—a cornerstone of the Standard Model of particle physics.

GÉANT engages with many such scientific projects. In this case, the huge amounts of data generated by the hadron collider experiments needed to be distributed to research institutes around the world for processing and storage. CERN had instituted a tier system, using the GÉANT backbone to carry the data. The particular technical report I was editing described a multidomain network monitoring and management service operating on the network.

We slowly went through the draft, and I asked Stefan questions based on the notes I had made.

"So, Stefan, if I can just put this in my own words, you're saying that the data on the network is distributed on the backbone down three levels—first to the main centre in Geneva, then to several centres across Europe and from there to the rest of the world?"

"Yes, Warren. That's correct."

Stefan's written and spoken English skills were not great—I realised that English was not his first language—so I encouraged him as much as I could. I allowed him time to speak, giving him the opportunity to reflect on his answers, occasionally asking questions to prompt him. By the end of the meeting I had a better understanding of what the report was about and who it was intended for. Some of the details of the technical implementation were still unclear but that didn't worry me too much at this stage—I would start by making a few small changes, and then after I had read the document several times and had a good understanding of the contents, I felt that I would be able to do a more extensive rewrite.

By the end of the week I had a report which I felt had far less repetition and jargon and was clearer to read and easier to understand. I submitted it to Catherine, the activity leader from the Spanish NREN, for approval.

"Good work, Warren!" she said to me smiling. Stefan was also happy and thanked me. I was relieved it had gone well. Some contributors can be quite touchy about having their work "rewritten," but Stefan had taken it in stride.

WEEKS 3–5: ATTEMPTS TO ENGAGE EUROPEAN RESEARCH COMMUNITIES

I was starting to settle in at ANT, making the most of the good weather and the beautiful surroundings. I had finally broken the ice with Jenny, who told me about the downstairs gym, which was free to all ANT staff, where she would disappear twice a week at lunchtime. I had also started using it, changing into shorts and a T-shirt for a 10-minute workout. Then I would take a leisurely jog out the building and down and along the scenic river Cam at Coe Fen, which was two minutes away. The daffodils and daisies were blooming, and the air was filled with the smell of fresh grass and cow manure from the cows that roamed freely along the meadows beside the Cam River.

Week 3 and I sat with David, one of the joint managing directors of ANT, reviewing the draft paper handed over to me by Craig, the task leader who had just broken his leg in a cycling accident and would be unavailable for the next few weeks. The paper was already late and needed to be submitted to the EC within the next two weeks.

Craig was prepared to continue working on the draft from home, but David would not have any of it. "He's either off sick or he's not!" he stated when I raised the suggestion that we involve Craig.

David was, one might say, a typical English gentleman. He had a wry sense of humour; he was reserved but with a brusque impatience and energy about him. He came across as nonconfrontational, but as I gradually came to see, he seemed to be used to getting his own way.

David had played a key role over the past 15 years in the running of ANT and the development of the GÉANT network. I had read the chapter he wrote in his book about the development of the European research network and been impressed by his contributions to GÉANT.

"This is just stream of consciousness stuff!" exclaimed David. He red-penned large sections of the document, commenting to me as he went. "This needs to go ... this is repetition ... this doesn't make sense...this is not accurate. This is just pure waffle."

Craig had left me with a long-winded document that rambled on. It read well enough, but David felt that the structure was not quite right and the document did not bring out the background information that he wanted to communicate. He explained this to me, discussed the structure he wanted, then handed me a copy of a book that explained the role and function of the scientific organisations within the organisation of the European Strategy Forum on Research Infrastructures (ESFRI).

"This is the way I want it done Warren!" he told me. "Read the book. It will give you some background."

I left David's office carrying his book and an armful of red-penned paper with David's hasty handwriting scrawled over it.

That evening, after work, I picked up the book and read through it. Believe it or not, I found it interesting learning about the different educational and scientific organisations across Europe, which GÉANT was trying to engage. It painted a really complex picture of how organisations across Europe were attempting to work together.

Part of the GÉANT project involved talking to members of the European scientific community, as well as other international organisations, to find out what they needed and how they could best be engaged. Often, initial engagement is through a pilot service and implementation of a full live service will then follow. The report I was working on was an attempt to describe to the EC reviewers this process of engagement with the European and international scientific communities.

A few days later I had an updated draft and was ready for a follow-up meeting with David. We sat in one of the meeting rooms, rather than his office, which meant that I could get the draft up on the overhead projector screen, so that we could go through it together.

"I read that book you gave me," I told him. "It was interesting reading. I think I have a better picture now of the scientific organisations around Europe. But I'm not sure how exactly this relates to the report we're drafting."

David sat back, joining the fingers of his hands. "We need to engage with all these groups. Some of the disciplines, such as telecommunications and networking, are better able to understand the benefits of GÉANT and are also more technically capable when it comes to connecting to it. But we are also trying to reach other groups that don't know much about us or have difficulty integrating their networks with GÉANT. We need to show them how to take advantage of the services we offer."

I thought about what he said, trying to digest this and build a mental picture of how it all fitted together.

Still confused, I said, "Okay, I understand the bit about engaging with scientific and research groups within the European community, but I don't understand how ESFRI is related to the GÉANT engagement process, or how the individual meetings Craig has been talking about are connected." For all I knew, ESFRI was an organisation founded by the European Union to develop the scientific integration of Europe and strengthen its international outreach.

David and Craig were both too close to their subject. They had become so familiar with the content and the context that they failed to fully grasp areas that someone encountering this report for the first time might struggle to understand. As the technical writer, totally new to their subject area, I had one advantage. Being remote from the subject, rather than immersed as a subject matter expert, made it easier for me to identify areas that were not clear or easy to understand by a nonexpert like myself who was reading the report for the first time.

"ESFRI is just one of the forums we are using to collect information from our European colleagues," David explained. "Craig has gone off and focussed mainly on his own activities, but that's only part of the picture of engagement. You see, the way it works is.... " David continued to describe how it all fit together.

Over the next two weeks I worked with David on the structure and contents, reshaping the report. Although David held a senior position at ANT, I was used to dealing with clients of different levels from previous roles at other companies. Although he could be insistent at times about what he wanted, I was able to relate to David as an equal and engage in an open dialogue; if I had treated him more deferentially, our communication might have been less effective and persuading him might have been less likely. Nevertheless, it still required some confidence in working with him, sometimes providing suggestions or guidance as to the structure of the draft where I felt it needed my input as a technical communicator or asking questions about and rewriting sections that were not clear.

For the main part, he agreed with my suggestions.

We finished the report and submitted it much to everyone's satisfaction. David hadn't said much, but I sensed in his general demeanour and engagement with me during our several meetings that he had enjoyed the review process and gained something out of it.

"Well done!" Catherine said to me warmly. "David is used to getting his own way and I thought you managed to deal with his requests and handle this deliverable very well."

As the technical communicator, knowing that I could help people develop, explain and communicate their ideas more effectively gave me a real sense of achievement. This was especially true in GÉANT's cross-cultural setting, where all the written reports must be clear and reader friendly, as they would be read by readers in other countries whose home language would not be English.

AN EMERGING EUROPEAN COMMUNITY

I had been at ANT for about two months, long enough to have a better understanding of how the company operated and some of the European concerns with which it had to deal. Many companies I had worked for in the past had featured ethnically and culturally diverse workforces, and I had often had to coordinate with offices abroad. Despite the diversity of these companies, the ethos and way of working tended to reflect the local conditions. So, for example, when working on a project for a large American organisation with a branch in the United Kingdom, the local work environment in Cambridge very much had a British approach and culture, with

British-style project managers and a British approach to work. The technical authors used U.K. spelling in user guides, and project managers adopted a nonassertive style that might have raised eyebrows across the Atlantic. What was unique about working for GÉANT was the fact that this was a truly cross-European endeavour. The team consisted of employees selected from the different European countries, and the project had a distinctly pan-European feel to it.

If you have read anything about the history of Europe, you can imagine that holding together and managing a project spanning so many interest groups and cultures can be challenging. For much of its history Europe has been a place of conflict and political upheaval, with constant wars waged between countries. World War II, only 60 years ago, has left a cultural legacy of distrust that is still active to this day. The formation of the European Economic Community (EEC) in 1957 brought Europe together in a common economic market and has gone a long way towards mending the wounds of the past. However, the legacy remains and is evidenced sometimes in cultural stereotyping and clashes that are still present. It is not just that people in different countries have a different vision of the future and their own national interests to protect but also that they perceive and react in different ways to events.

One example that illustrates this divergence is the adoption of the euro currency. Many of the central European countries, such as France, Germany, Spain, Italy and the Netherlands, have adopted the euro. However, countries further afield, such as the United Kingdom, have chosen not to opt in to the euro and have maintained their own currency, which provides more economic independence and allows countries outside the euro zone to set their own interest rates. What's more, the Second Gulf War exposed a rift in the viewpoints of EU countries, which had until then generally presented a united front on international policy.

But then again, with the emergence of the EU and the formation of the common European market, businesses across Europe are becoming more diverse. People are free to move across country borders and it is illegal to discriminate in the employment of staff based on their country of origin within the EU. Across Europe, there is also certain industry standardisation: We share a common technological infrastructure (e.g., using the same platforms and tools, such as mobile phones, the Internet and Microsoft products) and similar working practises and organisational structures. European legislation also has an impact that goes beyond borders and affects all companies across the EU. Directives from the EC and decisions made by the European Court of Justice are legally binding on all member nations, and this has had a huge impact on unifying local employment laws and policies, safety standards, human rights and other important matters across Europe. The history of GÉANT has very much reflected a general European theme—of postwar Europe working towards a common market and shared identity based on shared economic interests and security needs despite their divergence.

However, balanced against this strong drive towards unification was an equally strong spirit of independence in many of the European nations, a desire to do their own thing, in their own way, and not be totally reliant on Europe. In GÉANT this had manifested in the past in strong disagreements over the management and the future development of the network, from fundamental questions about the technology of the backbone infrastructure to the role of ANT as the operational management organisation. In my work as a technical communicator, I had also encountered differences of opinion as to whether our role and support were required, with one activity leader going as far as declining our offer to help.

WEEKS 6–8: DATA-SHARING REPORT

It was early May, the weather was warming up as we went into summer in England. I continued to enjoy long, lazy afternoon walks and jogs along the Cam River and other forested paths near the office. Having survived the first month, I was on to new and exciting things.

One of the problems with accessing and sharing data across the GÉANT network is that different federations/countries in Europe have implemented different technical standards for their networks. The GÉANT Authorisation Infrastructure for the research and education community (eduGAIN) service is one of the research activities of GÉANT which investigates ways in which federations can communicate with each other and share data via a secure and trusted interaction. The report I was writing was intended to describe this process and some of the technical solutions proposed for solving data sharing issues.

Valden, the subject matter expert, was based in Sweden. After communicating via e-mail for a little while, we arranged a meeting using Skype.

"So how's it going, Warren? How are you settling in at ANT?" Valden asked me once I had logged in and we had exchanged greetings. He had a pleasant voice and his command of spoken English was excellent, although I had noted some language problems in his written draft.

Valden was very much down to earth and had a friendly and open way of communicating which quickly put me at ease. We chatted for a while about various things before getting on to the draft.

"Valden, I've read the draft. I think it's a good start. I've suggested some changes. It could definitely do with a diagram in section 2 to explain the certificate process. We also need a better introduction to explain the background for the report."

We worked through the draft, and since I could see Valden on my Skype window, it helped to check if he was engaged and receptive to what I was saying. Being able to see and speak with Valden helped break the initial barriers and I was soon conversing with him more freely.

"Let's get together in London sometime," Valden suggested towards the end of the meeting. "I'm meeting with James next month and it would be good if you could meet up with us. I'll suggest it to James." James was the activity leader from the U.K. educational network organisation and one of the NREN partners. He agreed to the meeting.

A few weeks later and I was on my way to my first offsite meeting in London to discuss this project as well as prepare for the annual report, which was nearing the deadline for draft submission. The 50-minute journey on the fast train from Cambridge to London on a Monday morning was tranquil. It was after the morning rush hour, so the train was not packed. I sat by the window and watched the green English pastures and hills flowing by. I was looking forward to the meeting. There's a certain energy and excitement that comes from being taken out of your normal work surroundings and thrust into a novel environment, and I always enjoy a trip down to London, which is a great place to visit.

James's office was around the corner from Grays Inn. I grabbed a bite in the park before heading for the office. James met me at reception. A youthful looking English man with a full figure and smiling face, James looked relaxed and he quickly put me at ease with his laid-back, friendly manner. He took me to the meeting room and introduced me to Valden. I shook hands warmly, the ice already having been broken in our previous conversations.

We sat around the table in the conference room while I hooked up my laptop and put the report on screen for everyone to see. In-between the work, we sat and talked and I could sense by their tone of voice and demeanour that there was a good rapport between James and Valden and that they were pleased to meet me. The English and their Scandinavian colleagues, so it seems to me, often get on quite well—perhaps it is because we share the same gloomy weather and drinking habits. Or maybe it is language related as Scandinavians usually speak excellent English? At any event, we were soon all sitting back and gossiping informally, relaxed and sharing the occasional humorous comment. By the end of the meeting I felt that we had started the process of establishing trust and building a rapport, which I find so important in effectively working with others. I had gained something from our meeting—on both a personal and professional level—so it was well worth the time out.

That's not always the case in face-to-face meetings, though. I can often tell when there's a rapport by the fluency of the conversation and the relaxed look on people's faces. At the same time, I know there's a problem when the speech is strained and formal, with awkward silences and postures (as often happens when you're meeting with your boss's boss or someone who really doesn't want to meet you!).

Over the next week, Valden and I made good progress on the draft report; James signed it off and I submitted it to ANT's executive review board for approval. Comprised of senior members from the NREN organisations, this technical, policy and executive review board approves major reports that are for publication to the EC. Getting documentation through the review board was a lengthy and at times frustrating process for the subject matter experts, as it involved an iterative process of rewriting the reports to address issues identified by the reviewers.

Shaking his head, Walden exclaimed in frustration, "She just doesn't understand the subject!" He looked at me for support. The document had just come back from the executive review board and we were on a Skype session, going through Josette's comments. Josette was the head of the review board and a senior member of the French NREN.

It was awkward receiving negative comments from the reviewers, especially after we had spent a considerable amount of time working on the draft together. I looked at him, then asked neutrally, "Is there anything you think we can change to address her comments?"

"Well, the thing is, this is meant to be a technical report, and it's going to be read by technical reviewers. It's not going to make much sense or be of interest to anyone else! It's clear from her comments that she just doesn't understand the subject," Valden grimaced.

It was clear to me that Valden was quite adamant about the draft being accepted as it was. I had come to know Valden much better during the time we had been working on the draft. Although he was fairly outgoing and positive and able to engage effectively, I had picked up that he had a tendency towards mood swings and brooding. At any rate, I tried to respond in a calm voice.

"I think you're right—what we've written should be clear to any technical reader. If we put the information in section 2 after the introduction and then add explanations to the Appendix, that might make it slightly clearer? What do you think?"

Valden considered for a while, then nodded. "Okay, Warren," he sighed. "Thanks, I think that would be a good solution."

On a cross-cultural project, with different organisations and standards involved, I believed a review board to be a good idea. I had found it helps ensure consistency across reports and a consistently high level of technical accuracy and quality. However, it can be difficult to strike the right balance between effective review on the one hand, which finds errors and omissions and improves a draft, and excessive review on the other, which makes excessive or unreasonable requests and hampers the ability of writers to complete a document within scheduled deadlines.

In this case, I wondered if the executive review board had been a little overzealous in their review. However, we submitted the new draft with the changes I had worked out with Valden, and the review board approved it.

STARTING ANNUAL REPORT

About a month after I started at ANT Ivan informed me that I should get started preparing for the annual report. He sent me a link to the previous year's report and explained my role.

"You will need to contact each of the activity leaders and review their sections of the annual report, then put all the sections together and prepare it for publication to the EC."

This was actually more complicated than it sounded.

Each year, the task and activity leaders involved in GÉANT get together in a series of workshops to discuss and prepare for the annual report, a 150–200-page document summarizing the activities within GÉANT and their achievements over the previous 12 months. This report is intended for the EC and the NRENs. It is carefully scrutinised and reviewed by the EC and the NRENS to ensure that the project provides value for money and makes best use of the resources available.

It was mid-May when Kate, the English finance director, found her way to my desk. I was surprised to see her. She had been present at my job interview and was Ivan's boss. She had only been with ANT a few months and was constantly busy and often travelling. After asking me how I was settling in, she jumped straight in.

"I hope you're not planning to take any vacation time during July!" She stated emphatically.

July was the deadline for submission of the annual report.

"Don't worry. I'll be here," I reassured her but was slightly taken aback by her vigorous approach.

Ivan had not provided me with much guidance as to how to go about organising the compilation of the annual report. Without much guidance, I decided to use my own experience and approach to manage the drafting and compilation of the report.

Drafts and e-mails were going to be going back and forth all the time, from different contributors. If this wasn't clearly tracked, it could be chaos. To get a good handle on the project, what I needed was a good tracking system!

"What do you think of my project-tracking sheet?" I asked Jenny, passing a printed copy over to her.

The tracking sheet had not been asked for, but I knew from past experience that this would be a useful tool for keeping track of multiple drafts constantly being exchanged between reviewers, activity leaders and writers (see Figure 9.1).

Jenny glanced at it briefly. "It's okay," she replied nonchalantly, sliding the copy back, and then returned to her own work.

Like Jenny, Ivan and the other technical authors hadn't taken much interest in my efforts. They all were busy with their own activities and perhaps expected me to take the lead on my own projects. This is also something of a British cultural phenomenon—colleagues don't like to criticise or tell you how to do your job; it can be considered rude.

As the mid-June deadline for draft submission approached, the drafts began to trickle in slowly. When a draft was received, I marked the date received in my pro-

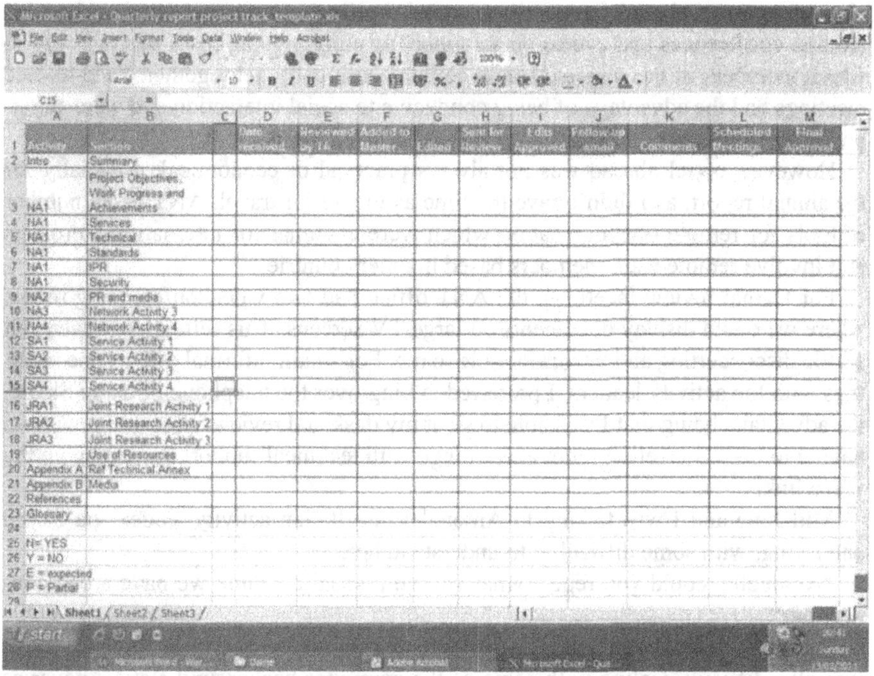

Figure 9.1   Draft tracking sheet. This figure shows the Excel sheet template used to track progress on the annual report.

ject-tracking sheet. I copied the contents of the document from the contributor to my master document.

I decided that it was important for me to have the received drafts reviewed promptly. I aimed to complete my initial review within two to three days of my having received the draft. As I needed to manage around 15 to 16 drafts from different contributors, this quick churn was important to keep the momentum going. Not all drafts were being received simultaneously, which was to my advantage in that it helped me to stagger the workload.

As I reviewed each draft, I marked up my revision comments and editorial changes/rewrites. Once changes were completed on the master document, I sent out a version of the full document to the contributor, indicating the section and page number where their content was located. I arranged for a follow-up call or meeting. I then filed away the contributor's version and marked on my tracking sheet that the content for their activity had been reviewed and sent back for review.

COMMUNICATING REMOTELY WITH ACTIVITY LEADERS

One advantage of working on a pan-European project is that travel distances between countries are not excessive (most locations are within a two- to three-hour flight) and flights are reasonably priced. It was not uncommon for some ANT employees to travel abroad frequently to attend meetings, conferences and events. Attending conferences and events on an annual or more regular basis was one way in which members of the dispersed team could occasionally get together. Face-to-face meetings had the advantage of being conducive to social interaction and were an opportunity to network and interact on a more informal basis.

However, travel abroad was not always practical or economical, especially for the annual report, as I didn't have the time available for travel. ANT had a number of tools for remote communication which were essential for international projects that involve remote team members based in other countries.

For formal review meetings, the ANT office had two videoconferencing rooms where we could display documents on large TV screens. This setting, however, felt a little disconcerting and uncomfortable to me. For small, informal one-to-one meetings with the activity leaders, I preferred talking over the telephone or using Skype, the advantage being that I was able to sit at my desk and review a document without having to book a meeting room and arrange with technical support for setting up the video link.

Mid-June and I was talking to Antonella, the Italian activity leader, via Skype and trying, with some difficulty, to understand her.

"Antonella, could you repeat what you've just said. I think we have a bad connection."

Antonella looked strained, struggling to communicate clearly in English, and it was also difficult dealing with some of the computer background static—never as clear as when you are speaking to someone face to face and can use their gestures and other cues to aid understanding.

I had reviewed her draft and found it difficult to understand and riddled with spelling and grammar mistakes. While I had similar problems to varying degrees with some other activity leaders, Antonella had the most difficulty.

I didn't like to put Antonella on the spot, so I decided it was best to keep the conversation brief.

"Antonella, let me make some changes. After I've done this, I will send the draft back to you for review. I think that would work best."

"Thank you, Warren," she replied gratefully. "If you look at the quarter report ... you have information ... you can read this also."

"Okay, thanks Antonella, I'll take a look at the quarterly report," I repeated. "It may have some information I can use."

I always remind myself, when listening to someone who has difficulty communicating due to language and cultural difference, what it was like for me many years ago as a new immigrant in Israel. I remember feeling frustrated at not really being understood and seen not as myself but rather as a stereotypical "immigrant" or "foreigner." It can be a very disempowering feeling. So now, listening to Antonella, I made a point of reminding myself that this was an intelligent, sophisticated lady who was making an effort to communicate in what, for her, was a foreign language.

After the meeting I had another attempt at updating the draft. Being able to track changes in the document gave me the confidence to red-pen and rewrite large sections electronically. But I was careful not to stray from the original meaning.

When working on the other reports with writers who lacked written English skills, I found that an iterative approach worked best. I read through Antonella's draft of the annual report, correcting basic errors in grammar and spelling as I went and making a note of any questions I had. Small steps. When a document is obscure and unclear, you have to chip away at it slowly, reshaping to reveal the meaning of the writer.

In the afternoon I made some more changes to her draft, then put it aside for the day, as there was only so much information my brain could process. The next day I reviewed the draft with fresh eyes, noting several errors that I had missed the first time. By now the draft was much easier to read.

Antonella's draft also needed structural changes, typical of drafts from some of the other contributors. She hadn't introduced key topics before discussing them, and linking sentences that would aid the flow from one idea to the next were missing. Luckily, the quarterly report provided more information. In essence, the annual report was a summary of the activities of the previous four quarters. The quarterly report had already been reviewed and approved, so I was able to plunder it for details.

JULY FIREWORKS

It was over a week after the submission deadline, and we still had not received all the drafts. The annual report had to be ready for publication by the end of July—a fixed deadline that had already been agreed to by ANT and couldn't be changed.

There we were, in a meeting room, discussing the progress of the annual report. Kate, our Irish secretary, who was helping Ivan with the year 1 preparations, threw her hands up in despair as she reviewed my tracking sheet.

"It's just awful!" she exclaimed in frustration, pounding the desk. "We've less than four weeks to submit this, and nearly half the reviewers haven't sent you their first drafts! How on earth are we going to get this done in time?"

Ivan surveyed her calmly, showing no emotion. "I will raise it at our next team leader conference call," he stated simply.

"Small steps, Kate," I added. "We can't solve everything at once. One step at a time."

I had been quite laid back and friendly towards the activity leaders, I thought as I was walking out of the meeting room, but it was time now to turn the pressure up a notch with some of the tardy contributors. And I knew who I needed to confront first.

"I've been waiting for over two weeks now for Hans's draft, and each time he promises me it's going to be a few more days!" I grumbled. In the kitchen, I was complaining to Jenny.

The rich smell of ground filter coffee wafted through the air. Jenny was an expert at preparing filter coffee—much better than the standard stuff that came out of the tin.

She poured coffee into a mug and handed it to me. During the past month there had been a marked change in Jenny's attitude towards me—she had opened up and I found a warm, accepting person who was easy to get on with.

"Thanks Jenny!" I replied, taking a sip of coffee.

"Ooh, don't tell me about Hans!" Jenny commiserated. "You're lucky you don't have to work with him on a regular basis! Sometimes, he can be just impossible!"

Hans was the activity leader of one of the areas that Jenny covered. He was a senior figure in the German NREN organisation.

She sighed, sipped her coffee, then put it down with a soft bang on the table. She continued, "He never responds to my e-mails, and he always submits drafts at the last minute!"

"Well, I'm not going to put up with that," I stated. "I don't care how important he thinks he is!"

Despite my brash words, I realised that when dealing with tardy activity leaders like Hans I had to tread cautiously. Many were senior members in their own NREN organisations. They did not report directly to ANT and a few did not appreciate the role of ANT, or the technical communicators for that matter, in the project. Nevertheless, I felt that I needed to be polite but firm.

It was Monday morning, the first week in July, and I was on the phone with Hans, attempting to resolve the issue of his late response. "Hi Hans, thanks for taking my call," I started. "I hope all's well with you—I'd really like to discuss your draft."

"Okay, go ahead please," Hans responded. "Please be quick. I have a meeting soon."

"I haven't had any response to the e-mail messages I've sent you. I really need to know if you'll be able to get your draft of your section of the annual report to me by this Wednesday," I said, getting straight to the point.

Hans took a moment to respond. "I don't think so. I am sorry. I don't have time now. I have so many other deadlines."

I paused.

"Hans, I'm afraid that I may not have time to include it in the annual report if we don't have it by this Wednesday. I'm sorry, but the deadline was two weeks ago, and we've just run out of time. I wish there was more time available. Would you mind if we included your section in one of the annexes, which will be available shortly after we've published the annual report?"

There was silence on the other side for a moment. I sensed Hans was somewhat taken aback. "Okay, I will see what I can do," he responded gruffly.

It did the trick. I had my response by Wednesday.

LAST HURDLE

The end of July deadline for publication was fast approaching, and I was working hard to make sure that we met it. The majority of sections had been compiled, reviewed and redrafted. There were a few straggling items that still needed completing.

The activity leaders were all visiting ANT in the week leading up to the presentation of the annual report to the EC reviewers, so I made sure that those I needed to speak to had pencilled in some time to sit with me and go through any remaining questions or issues about their section.

One week later and the report was ready—all sections had been edited, compiled, reviewed, rewritten and signed off by the activity leaders. It was now ready to go to the executive review board.

The annual report, because of its high profile and importance to the project, would be subject to intense review by the executive committee, which, as described earlier, was a group of senior executives selected from the NREN organisations responsible for the decision making and management of GÉANT. And the board was headed by Josette from the French NREN.

Realising how important this review was going to be, I carefully crafted an e-mail, attached the final report in PDF format and sent it out to the executive reviewers. I also wanted to make sure that I adequately collected and addressed each of the reviewer's comments when they came back, so I put together the comment-tracking sheet shown in Figure 9.2.

A few days later, review comments began to come back from the executives on the review board. Most were reasonable. Josette had responded in her usual, blunt manner: "I've reviewed this document and it's given me a GÉANT (giant) headache!" she stated in her response.

I sighed, not appreciating her humour at the expense of the team who had spent a great deal of time putting together the report and wondering whether any of the activity leaders might take exception to her comments. I reviewed her comments, which luckily were not that extensive, then contacted those activity leaders with whom I needed to further discuss their sections.

Some of Josette's comments had to do with Hans's section so I connected with him a few days later on Skype to see what we could do.

Hans eyed me suspiciously through his thick-rimmed glasses on the monitor. "For some reason they have it in for my activity!" he grumbled. "Look at this com-

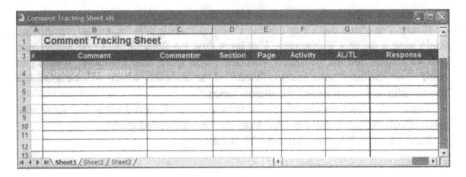

Figure 9.2 Tracking sheet template. This figure shows the Excel sheet template used to track reviewer comments on the annual report.

ment—they are not fair! We have done just what the other activities have done, and can you see how they are picking on us!"

The comment Hans was referring to had come from Josette. She wanted him to restructure his introduction to provide more of a background explanation about his activity, and she had made a number of specific comments about this. However, many of the other activity leaders had adopted an approach similar to Hans'—keeping the introductory section short and based on the information from the quarterly report and technical annex—so Hans felt singled out and picked on.

I sensed his frustration. It was obvious from previous meetings that he and Josette didn't always see eye to eye. Having talked to Jenny and Ivan, I understood that Hans had managed to get the backs up of some of the project team. Perhaps the reviewers were coming down harder on him because he had been difficult in the past? Being fairly new to the project, I had no real way of knowing what he had done to offend and could only base my assumptions on what I had heard from others in the team and what I had observed of his behaviour during the times we met. It seemed that his approach was at times confrontational, and I thought that was bound to lead to problems and cause resentment from others in the team.

"This is all politics!" Hans complained.

I raised an eyebrow. So who's playing politics now? I thought to myself.

"Okay, I understand what you're saying, Hans," I responded, "but let's just see if we can't at least address their comments. If we can give a reasonable response, then at least the reviewers won't feel as though we've just ignored them. What do you think?"

As far as I was concerned, there was no point in alienating the reviewers by being dismissive of their comments, taking sides or responding defensively.

Hans grudgingly concurred. However, when I got back his responses a few days later, he had suggested only a few changes but had responded to many of Josette's comments in a dismissive and defensive manner.

I wrote back to him and made some suggestions as to how we could deal with some of the comments in a more constructive manner. He didn't respond.

I shrugged my shoulders and put the updated draft with Hans's changes, including my own amendments, out for review, making a note of each of Hans's responses in my review-tracking sheet. There was only so much I could do to help or suggest changes; at the end of the day, it was up to the activity leaders to decide how they wanted to respond.

For me, the clash between Josette and Hans, both senior members of independent NREN organisations, reflected a certain megalomania and egotism that possibly wouldn't have been tolerated if that behaviour had been exhibited by less senior members in the project. I was not sure how much of this had to do with the personalities involved, how much represented a clash between cultures and how much reflected a different national view of the way in which the project was being run.

I tried, however, not to fall into an easy, stereotypical perception of Hans as the uncommunicative, reserved German and Josette as the arrogant French lady. This would have been unfair, both to them and to the rich cultures from which they came. Whatever the reason for their differences, it was in my own interest to ensure that the draft would get through to publication to the satisfaction of all parties, without alienating any of the key stakeholders who I would need to work with at a later stage. Sometimes, as the technical communicator, you are in a position to be able to help mediate and negotiate between parties that don't see eye to eye. It requires remaining as objective and neutral as possible; nevertheless, there's only so much you can do.

AFTERNOON TEA BREAK

I found the kitchen a good place to sit and relax in the afternoon with a warm cup of tea and piece of chocolate. The windows from the kitchen gave a good view of Cambridge, with the train station visible about a quarter mile away.

It was quiet at this time of the afternoon, and I was alone in the kitchen. I was gazing out the window when I heard the sound of footsteps and the kettle being switched on. I turned around to see Catherine, the Spanish activity leader.

She gave me a quick smile, "Hi Warren."

"Hi," I responded with a smile.

She put a tea bag into her cup and then added water from the boiling kettle. "You seem lost in thought. How's the annual report going?" she asked.

"It's going okay. The report's almost there. Just one or two small issues with some of the review comments I've had to chase up."

"Oh," Catherine paused. "Hope no one is giving you any trouble?"

"Well, not exactly."

I had a rapport with Catherine, who was always approachable. I was soon explaining to her the incident with Hans and his reaction to Josette's comments.

"I wonder if Hans and Josette clash because they are both such dominant people," I offered as a suggestion. "Josette comes across as quite arrogant—I don't think her approach goes down very well with Hans."

"Mmm ... I admire Josette," Catherine responded. "She's just so direct and speaks her mind and won't let anyone intimidate her."

I was quite surprised upon hearing this. Then I realized something I had not thought about before: Although I personally found Josette's approach to be somewhat arrogant, she was very much admired by others in the team for the very traits that I found distasteful! As a female leader in a more male-dominated working culture, independence of thought and confidence to speak your own mind can be considered a trait worth admiring even if that might come across as arrogant to some others in the team.

REPORT SIGNED OFF

It was two days after my meeting with Hans. I sat in the boardroom next to Ivan. The boardroom, with space for up to 40 seats, looked somewhat empty with just me and Ivan. We sat facing the large TV screen, which had remote link-ups to other offices in Europe. Josette was visible on the screen. We were here to address her comments after discussing them with the activity leaders and Ivan. We felt that a few of her comments were unclear and some of them suggested she had misunderstood or missed the point.

Ivan introduced us (this was the first time I had met Josette), explained where we were with the report and then handed it over to me.

"Thanks Ivan," I responded, "and good to speak to you Josette. If you don't mind, I'd like to run through your comments and then discuss how we've implemented them. Should we start at the beginning?"

We all had a version of the report open in front of us. Track changes had been put on to show where changes had been implemented. At the same time, I had a copy of my reviewer's tracking sheet open (see Figure 9.2) so that I could run through the comments and indicate how we had responded. Josette nodded and I began.

It was a long document and there were about 40 comments to go through. The comments relating to typos or misspellings were self-explanatory and quickly dealt with. Where Josette had asked for more detail and clarification about a point made in the report, I explained what had been done by the activity leader to address her concerns and we reviewed their response, highlighted with track changes. For most of these she was satisfied or suggested some minor wording changes. A few required further discussion and clarification around what she had meant by her comments. I made some notes for follow-up.

We came to Hans's section, and I showed the explanation that Hans had given regarding her comments. "Josette, I suggest that as we haven't adopted that approach for the other sections, we leave it as is." I said diplomatically.

Josette considered for a moment, then nodded. "Okay Warren, I am happy with your suggestion."

I cannot say I wasn't surprised at how Josette let this go, but I was, for the most part, just happy that she did.

We spent the next 20 minutes going through the remaining sections. Ivan allowed me to do most of the talking. At the end, he thanked Josette for the time and ended the video connection.

Ivan shut off the TV monitor and turned to me. "How do you think that went?"

I shrugged my shoulders. "It seemed to go okay," I responded. "There were no major issues, and she was happy with Hans's section."

Ivan nodded cautiously, considering for a moment. "Yes, I agree."

I left the room, thankful that most of the issues had been resolved. The conversation I had a few days before with Catherine, who made the point that she admired Josette for her direct, no-nonsense approach, was still fresh in my mind. And then I thought of Ivan, who was probably the opposite of Josette, sitting like a closed book during the meeting. He always kept his cards to himself and never communicated what he was really thinking—a trait I found frustrating. However, having lived in an ex-communist country, he might have found that trait very important for survival. Being able to be discrete without disclosing your position was certainly also a useful trait to have in Ivan's current position as the ANT project coordinator.

A few days after the review session with Josette, I sent through the final version. She was satisfied that all her comments had been addressed. The annual report was signed off by the EC reviewers with some positive comments, and David sent out an e-mail to the team, thanking them all for their hard work and contributions.

WORLD CUP FEVER

It was the time of the FIFA 2010 football World Cup in South Africa, and everybody in the office was abuzz with excitement. The greats of Europe—England, France, Spain, Germany, Holland, Italy—were playing, together with representatives from other nations. GÉANT, being such a diverse project, had many staff whose home teams were represented. There was much friendly banter going around the office— who would win and which teams would get through to the next round. The activity leaders from across Europe were all present for the presentation of the annual report to a group of senior officials from the EC, and they took part in the friendly banter.

It was a source of friendly rivalry, taken with a good sense of humour in the office, and people whose teams got through to the next stage were proud and pleased. The excitement in the air was palpable.

In the midst of this fever, something hit me. I thought about how the development of the GÉANT network was a positive example of European countries working together, trying to overcome their political and cultural differences in order to establish a network that would be of benefit to all parties involved.

Yes, people in Europe identified with their home cultures and teams and had a distinct identity as German, French, English, and so on, but there was also a strong sense of a shared community and mutual respect amongst the members of our project team, of people able to work together and communicate effectively to achieve a shared vision. The openness of the team leaders and their willingness to cooperate had been an essential element in the project's success. As a technical communicator, I saw my role as being that of a facilitator and a means of bridging the differences— of listening to all sides and helping them to communicate their ideas more clearly and effectively.

I was only with ANT for a short period, but the experience of working on the GÉANT project with its diverse, multicultural working environment had emphasised to me just how interdependent this modern society had become.

RECOMMENDED READINGS

1. The following sources discuss the development and practice of technical communication in Europe:

   - J. Kirkman, "From chore to profession: How technical communication in the United Kingdom has changed over the past twenty-five years," *Journal of Technical Writing and Communication,* vol. 26, no. 2, pp. 147–154, 1996.
   - M. Krause, "Professional education of technical writers in Germany," *Technical Communication,* vol. 42, no. 1, pp. 173–176, 1995.
   - M. Austin, "Documentation and that directive," *The Communicator NS,* vol. 3, no. 9, pp. 2–3, 1992.
   - P. Hunt, "The teaching of technical communication in Europe: A report from Britain," *Technical Communication Quarterly,* vol. 2, no. 3, pp. 319–330, Summer 1993.
   - S. Nilsson, "Understandability of narratives in annual reports," *Journal of Technical Writing and Communication,* vol. 27, no. 4, pp. 361–384, 1997.
   - F. Salager-Meyer, "Debate-creating vs. accounting references in French medical journals," *Technical Communication Quarterly,* vol. 9, no. 3, pp. 291–310, 2000.
   - C. Taylor, "Information design: A European perspective," *Technical Communication,* vol. 47, no. 2, pp. 167–168, 2000.

2. The following sources provide information on European cultures and how cultural preferences may affect technical communication:

   - J. M. Ulijn, "Translating the culture of technical documents: Some experimental evidence," in *International Dimensions of Technical Communication,* D. C. Andrews, Ed. Arlington, VA: Society for Technical Communication, 1996, pp. 69–86.
   - I. Vankevič, "The main skills and competences of multilingualism in the context of the European Union." *LIMES,* vol. 3, no. 1, pp. 55–66, 2010.
   - J. M. Ulijn and K. St.Amant, "Mutual intercultural perception: How does it affect technical communication? Some data from China, the Netherlands, Germany, France, and Italy," *Technical Communication,* vol. 47, no. 2, pp. 220–237, 2000.
   - Y. Wang and D. Wang, "Cultural contexts in technical communication: A study of Chinese and German automobile literature," *Technical Communication,* vol. 56, no. 1, pp. 39–50, 2009.

3. This story discussed the challenges of managing and coordinating documentation projects in a global workplace. The following source provides more information on this topic:

   - D. Voss and M. Flammia, "Ethical and intercultural challenges for technical communicators and managers in a shrinking global marketplace," *Technical Communication,* vol. 54, no. 1, pp. 72–87, 2007.

DISCUSSION QUESTIONS

1. Warren did not quite identify his own cultural background. What do you think it might be? How do you think this might have influenced his perception of the various people with whom he worked?

2. Thinking of the European examples in this story, what distinct European cultures could you identify? Do you think there is a sense in the story of a shared European culture emerging? Can you think of other examples of cultures and countries moving closer together? What are some of the reasons for this?

3. Warren talks about initial difficulty breaking the ice with the other technical writers and clashes between some of the project leaders. How much of this do you think is related to "cultural differences" and how much is due to individual personalities? How would you differentiate between "culture" and "personality"?

4. Warren refers to cultural stereotypes. Can you think of some other examples of cultural stereotypes and discuss how such stereotypes might have arisen?

5. Warren comments on how he dealt with contributors who did not speak English as their first language or who had poor written and spoken communication skills. What were some of the strategies he used to deal with this? Can you think of other strategies he might have used?

6. What do you think was the key role the technical writer played and the benefits he brought to the project in this story?

CHAPTER 10

ARE CHINESE DOCUMENTS READY FOR GLOBAL AUDIENCES?

Melanie G. Flanders

Melanie G. Flanders has been a technical editor and writer for over 30 years. She has worked in numerous industries, including computer and telecommunications hardware and software, engineering, construction, medical, banking, retailing, oil and gas, human resources, publishing, and education. Flanders also has over 13 years' experience as a corporate trainer, teaching courses in information design, business and technical communication, and FrameMaker.

Before relocating to Nanjing, China, in April 2005, Flanders lived in Houston, Texas, for 15 years. In 1998, she started KnowledgeMasters, Inc., a documentation services, training, and consulting company. From 2001 until 2005, she taught writing and electronic publishing courses at the University of Houston—Downtown and Houston Community College. Flanders has been a longstanding member of the Society for Technical Communication (STC), and in 2005, her colleagues conferred her as an associate fellow.

Since coming to Nanjing, she has worked at a software company as a technical writer, taught spoken English at Nanjing University of Aeronautics and Astronautics (NUAA) and computer science and spoken English courses at Nanjing University of Posts and Telecommunications (NUPT), and done corporate training in business English and cross-cultural communication. She currently works for Sunyu Translation Co. as the only editor who is a native English speaker.

CHAPTER SYNOPSIS

Ever wondered what it is like for a foreigner to live and work in China and what it takes to do technical communication work at a Chinese company? This story

Negotiating Cultural Encounters. Edited by Han Yu and Gerald Savage
Copyright © 2013 The Institute of Electrical and Electronics Engineers, Inc.

gives you some ideas. Melanie, an entrepreneur, teacher, and technical editor from the United States, tells her story living in a second-tier Chinese city, Nanjing, working for a rising telecommunications company. The company hired Melanie to "Americanize" product documentation for its growing global customer base. The subject matter experts at the company would write the documentation in Chinese, a group of translators would translate it into English, and Melanie, together with a small group of native English speakers, would edit the translation to ensure it adhered to standard American English. Additionally, Melanie would note recurring errors in the English translations to provide training sessions to translators.

In this chapter, Melanie shares how she established contact with the company, how she impressed them well enough to be hired, what the highs and lows of her job were, and what her thoughts are regarding the global-readiness of Chinese technical documents. Along the way, we get a glimpse of what it is like to teach in a Chinese university.

In early 2005, I arrived in Nanjing, China, to work as a senior technical writer for a software company. Nanjing is known and respected for the number of universities and multinational companies that it has. Unlike Beijing, Shanghai, Guangzhou, and Shenzhen, which are "first-tier" cities in China, Nanjing is a "second-tier" city. It is not as large or as developed as the first-tier cities, but it is still a major contributor to China's economy and is much more modern than other places in China. Other second-tier Chinese cities include places like Hangzhou, Xi'an, and E'erduosi.

LISA WORKSHOP

In April 2006, the Localization Industry Standards Association (LISA) held a three-day forum and workshop in Shanghai, China. As part of the conference, LISA invited me to present a one-day workshop entitled "Writing Technical Documentation: A Primer for Non-English Speakers." Being the entrepreneur that I am, I said yes.

The workshop was easy to prepare. I had presented similar workshops during the eight years that I owned KnowledgeMasters in Houston, Texas. Having lived in China long enough to have some idea of the common mistakes that Chinese English speakers made when they spoke and wrote English, I felt I knew how to tailor the material for my prospective audience.

To my delight, there were 13 participants. Two had come all the way from Singapore, 6 were from two major telecommunications companies, and the remaining attendees were from various translation companies in China. All were native Chinese speakers.

The hotel provided us with a sumptuous buffet lunch after the morning workshop. I was the last to go through the line and looked at the tables to see which one was open to having me join them. One of the telecommunications companies, RGS Technologies (all company and character names used are pseudonyms), which was based in Shenzhen with several offices throughout China, including Nanjing, had three people at the table and they smiled as I approached.

"May I join you?" I asked with a smile.

"Of course, please sit down," one of them said.

I sat down and we exchanged a few pleasantries. I asked them some questions to learn more about them, their company, and their respective roles in the company. They asked me how I liked living in China. I was very careful not to try to do any kind of sales pitch, as that would violate a Chinese protocol. In China, people want to get to know you first before they are interested in doing business with you. This is known as building *guanxi*. *Guanxi* is similar to networking but is more complex. It involves more than just staying in touch with people and then calling on them when you need something. (In our professional group in the United States, we always knew when certain people were looking for a job because that was the only time they ever attended meetings or called or e-mailed people in our group.) With *guanxi*, you help people with what they need and thus make friends. To keep the relationship going, you continue to exchange favors. You need to establish trust and credibility with colleagues and potential clients first before anything else can happen. Without *guanxi*, it is extremely difficult to do business in China.

All three men introduced themselves with their English names rather than their Chinese names. It is common for English-speaking Chinese to use English names, but since they all use their Chinese names in the workplace and are known by their Chinese names, I have learned to ask someone's Chinese name as well. Will was the senior member of the group. He was based in Shenzhen but was from a city near Nanjing. Richard was based in Nanjing and reported to Will. Raymond also reported to Will and was based in Shenzhen.

Midway through the meal, Richard smiled and said, "We were wondering whether you might be interested in doing some cooperation with our company. We are looking for ways to improve the quality of the English in our documentation."

"I would be delighted," I replied, thinking to myself that I had apparently made an impression on this group during the morning workshop. "How do you think it should be improved?" I asked.

"Someone will contact you to discuss it when we return to Nanjing," said Richard.

In the workshop after lunch, I focused on Chinese-specific issues with writing in English and allowed ample time for questions and answers, which was a good tactic, as the participants asked numerous questions. Their questions ranged from points of grammar and phrasing to how to develop documentation plans.

I used examples that compared Chinese and English, so that my participants could easily "get" what I was trying to teach them. For example, verbs, which are essential to English sentences, are not always required in Chinese. Instead of saying "I am happy," in Chinese, people say, "I happy." When I use examples such as this one, people are more easily able to understand the need for verbs in English. Mandarin does not have all the verb conjugations that English has. To demonstrate use of the past tense, I compared past tense with the Chinese particle 了 (le), which is often used to indicate that something has already happened.

We also spent nearly an hour discussing the documentation process. They were unfamiliar with the concept of documentation plans, so I showed them some sample plans so that they could see how the plan could range from a simple one-page plan to a whole document, depending on the complexity of a documentation project.

I met several other people during the forum, including several Europeans who were interested in the fact that I had been living in China. I made several business contacts, learned some things along the way, and had a very productive three days. I felt I had set the stage for "cooperation" with several companies.

In China, one either participates in "cooperation" or "collaboration." This is important to know, because if you call yourself a "consultant," you will not get hired. Chinese do not hire "consultants" because they do not believe in paying for advice. You must be able to provide a service for which they can then hire you to cooperate with or collaborate on some project. You are still serving as a consultant; however, to market your services, you must necessarily shift to meet these business perceptions.

RGS TECHNOLOGIES WORKSHOP

Three months after the forum, Richard contacted me and invited me to have lunch with him. I met him at his office and then we walked to a nearby western-style coffee house/restaurant. I knew that the purpose of this lunch would simply be to get to know each other better, so I let Richard lead the conversation. Again, I was very careful not to launch a sales pitch because the purpose of the lunch was to build *guanxi*.

"How long have you been with RGS Technologies?" I asked.

"About 12 years," Richard replied. "I used to be a teacher before that."

"Ah," I said. "And what did you teach?"

"I taught English at the university," he replied.

"Did you enjoy it?"

"It was OK," he said, "but I enjoy my work now much more. And, of course, the salary is much better." He smiled.

"Ah. I taught business and technical writing for several years in the United States," I said. "I taught it in the classroom and also as an online course. I also taught how to write manuals and electronic publishing."

"So you have a lot of experience teaching?" asked Richard.

"I have taught at universities part time for over five years and my company did training for several companies in the United States. I have also made two trips to Taiwan to do some training for my current employer."

We had a few more exchanges, and lunch ended with Richard telling me that he would be in contact with me again soon.

About six weeks later, we met for lunch again. This time, Richard brought a list of things that RGS Technologies wanted me to be able to provide and asked me to prepare a proposal that would address how I would deal with these items. We discussed the items briefly so that I could get clarification on some of them. I would be working with the translation department. I asked him when he wanted to see the proposal.

"Oh, no hurry," Richard said. "Sometime in the next month will be fine."

I went home and prepared a proposal that addressed their concerns. I also submitted a price for delivering a two-day workshop. I had to submit it to his boss, Will, because Richard was not authorized to receive external e-mail at that time.

At the third lunch meeting, Richard asked me to sign a nondisclosure agreement that would allow him to deliver a CD to me that contained some of their manuals so that I could see what their documents looked like. A couple of weeks later, I went to the office and picked up the CD.

By this time, we had agreed that the two-day workshop would focus on grammar rather than content, and I would base the workshop content on grammar issues that their translators were having and draw on the documents that they had given me for examples.

Richard invited me to dinner and we shared two bottles of beer. I knew that I was making progress; that is, we were establishing *guanxi*. After the dinner, I began to prepare the workshop material. We had a second dinner and set the workshop date for late November. I was to have about 90 participants, and I would teach the workshop in Shenzhen, the company's headquarters.

In October 2006, I had left the software company where I had been employed for 18 months and was working for a U.S. company based out of Dallas as its sole employee in China. In November, I went to Manila (the Philippines) for 10 days to meet the U.S. company's team there and then flew to Shenzhen to meet my clients. It had taken 6 months from the time I first met Richard until I actually delivered the workshop.

The first day of the workshop, I had 96 attendees. I introduced myself in Chinese and then started the workshop in English. I repeated much of the material I had covered in the workshop in Shanghai, inserting some specific examples for my audience.

On the second day, attendance was lower because many of the translators couldn't afford to take two days' time away from their work. I focused on the examples that I had prepared using the manuals from the CD that they had given me.

We had a lengthy discussion and the translators asked really good questions.

"Why did you change 'The cable can be connected to the serial port' to 'Connect the cable to the serial port'?" asked one of the translators. "Isn't that rude?"

"No," I replied. "'The cable can be connected to the serial port' simply states that it is possible to connect the cable to the serial port. The sentence does not instruct you to connect the cable. This is a procedure. With procedures, we use the imperative, or implied second person, to communicate instructions. It is not considered rude. It is direct language that tells someone what to do.''

"Why did you change 'The cable may be loosely connected. If this is the case, tighten the posts that screw into the serial port' to 'If the cable is loose, tighten the screws on either side of the serial port'?"

"'If' indicates a condition, so 'If this is the case ...' is redundant. By placing the word 'if' in front of the first sentence, you have now stated the condition. If this condition is not true, the reader can stop reading right there. If the condition is true, the reader immediately knows what needs to be done. I changed the wording about the location of the screws because the previous wording is confusing. Everyone in the target audience knows what a screw is and that you can only do two things with one—tighten, or screw, it, and loosen, or unscrew, it. The previous wording implies that the screws are actually in the port, which they are not. They are on either side of it. With the previous wording, you would have credibility issues with your audience because they would think that you didn't understand the equipment."

"Oh, now I understand. Thank you for clarifying that for us."

The workshop ended well, and that night, we had a celebratory dinner in typical Chinese style to discuss the feedback that we had received. Celebratory dinners are always held around round tables in China. Typically, a specially folded napkin will denote the place of honor and guests are seated around the table in proximity to the honored guest according to their importance. I always wait to be told where to sit. At this dinner, I was at the place of honor. Will and Richard sat on either side of me, with Raymond next to Will. The other eight people who helped put the workshop together were seated at the remaining places. We had a sumptuous meal with about 15 dishes and drank several toasts. I returned to Nanjing the next afternoon.

LESSONS LEARNED FROM TEACHING AT A UNIVERSITY

For nearly a year, nothing further came from the discussions we had during the dinner. In October 2007, I started teaching English at Nanjing University of Aeronautics and Astronautics (NUAA). During the two years, I discovered how Chinese students learn English, the pronunciation difficulties they have, and the way that they put English words together to form ideas. For example, "Welcome to come to Nanjing" means "Welcome to Nanjing," "Attention! Your head!" means "Watch your head," "Closing the door in your hand" means "Do not put your hand between the doors when it is closing," and "Carry out implementation management of the project" means "Manage the project's implementation." Other problematic phrases occur when the intended meaning is different from what is stated; for example, "Eat lunch by yourself" means "Lunch on your own," and "Prepare the application in time" means "Complete the application in a timely manner (or within a specified period)." Some Chinese translations also tend towards using gerunds instead of nouns or verbs; for example, "No hands touching" and "inductive washing" mean "Automatic faucet" and "No grass walking" means "Keep off the grass." From these insights I gained, I developed pronunciation exercises and lists of phrases to help my students.

GETTING TO KNOW RGS TECHNOLOGIES

I stayed in touch with Richard through text messages and an occasional e-mail message. In August 2009, Richard asked whether I could come to his company to help out with some templates they were trying to "Americanize." I said yes.

I spent most of the day with Ren Hao Li, or Harry, one of the senior translators. We got through the templates much more quickly than either of us expected, and as we went through each one, Harry would occasionally ask the reason for the suggested wording changes and I would explain.

At the end of our session, Harry said, "Thank you! I learned so much from you today. I look forward to cooperating with you again."

Richard then came in to talk to me. "Do you think you can come in one or two days a month to help us out with some documents?"

"Sure. That would be no problem."

I didn't hear anything from Richard for nearly three months. (It turned out he had to wait for his budget to come through.) Then, in mid-November, he called me. "We want to know if you can come in one or two days a week. When can you start?"

My first day was December 7, 2009. My primary tasks were to review documents for content and to "Americanize" them. They asked me to note any recurring errors and phrasing issues and to provide monthly training sessions to the 30 or so translators in the department.

RGS Technologies was, and still is, in the process of becoming a global operation. Its initial forays outside of China were developing countries in Africa, the Middle East, and Eastern Europe. Because Chinese employees did most of the installation and after-sales support, the quality of the documentation was considered unimportant. Now that the company has moved into Western Europe, North America, and South America, where the support engineers are native to their respective countries, it has become important to provide better quality documentation. The company has also landed some major contracts to supply mobile communications switching equipment, handsets, and software products and services to companies in the United States, and improving the quality of the documentation has become a necessity. The company's ultimate goal is to be able to create all its source documentation in English to facilitate translation into the many languages in which it now needs to distribute its documentation and produce "world-class" documentation.

At RGS Technologies, Chinese engineers and software developers, rather than technical writers, write documentation in Chinese. They do not have any Chinese technical writers because they cannot find any to hire. The universities are not producing graduates who have studied technical communication because it is not available as a course of study or as a major. Technical writing as we know it in the West is still in its infancy here, but many companies in China have recognized the need for technical communicators and are pressing universities to create technical writing programs as part of their curriculum. However, it is very difficult to institute change in Chinese universities, and it may be years before we actually see comprehensive technical communication programs as a common offering.

To complicate matters even more, many Chinese people think of technical documentation as something that is written in English, not Chinese. As one Chinese colleague at RGS Technologies told me, "Direct, clear, concise technical documentation is impossible in Chinese—it is against our culture."

"But that is exactly why we need to teach technical documentation in Chinese, and not just in English," I replied. "Because it does mean a whole new way of thinking. Technical documentation is not just something in English—it applies to technology worldwide."

RGS Technologies is in the mode that most U.S. companies were in 20 years ago and some companies are still in today; the management position is that if you can speak a language, you can write in it. Sound familiar? As we all know all too well, some people are good writers, some are mediocre, and others are just plain terrible. This is a fact that transcends culture and language. Unfortunately, at RGS Technologies, the Chinese writers' inability to express themselves well when writing in Chinese makes the translators' jobs difficult because they don't have good source text to translate into English. Sometimes, the translators have to ask the writers for clarification on the intended meaning so they can translate it into English.

Many of the translators were formerly English teachers. Only two have had the opportunity to spend any time abroad. Most are fairly proficient at translating written Chinese into English, although their translations contain the same nonstandard phrases I had observed in the work of my students at NUAA. Some are fairly proficient in speaking English, but a few have difficulty expressing themselves verbally, mostly because they don't practice speaking English—they either haven't had the opportunity or are too timid to speak because they think their pronunciation is poor. Their backgrounds in English are primarily literary, not technical, so until they learn about the products, they often have difficulty in choosing the right vocabulary.

RGS Technologies uses automated translation tools such as Trados to facilitate translation. The automated tools have a database of words, expressions, phrases, and sentences that have previously been translated. The translator can choose to use these items to expedite the translation process, but the tools work most efficiently when the writers use consistent phrasing. Unfortunately, most Chinese writers are not very consistent in the way they present information, so the translators are not always able to use the automated tools and must instead do a manual translation. If the Chinese phrase is something they feel will be translated often in the future, the translators can add the phrase to the translation tool's database. Most of the time, the translation tools are very useful, but once in a while the result can be ludicrous, as is shown in the following letter that I received when looking at apartments. The letter is a Google® translation that someone who did not speak English prepared.

Concerning the community
    1: Our community environment is beautiful, there are a lake and very big gardens. Still have the tennis court, gym. The community west front door contain bank, the convenience store.
    2: The community west front door street is having the biggest pet of Nanjing hospital in front.
    3: The transportation convenience, arrive the car distance that the large supermarket and new business district of streets only have 10 minutes.
    4: You can let send to everyday of the milk work to send to milk the family, you can also arrange everyday of the hour work or sweep hygiene for your residence every week.
    5: This community likes the person of the pet a lot of, everyday morn-

ing and evening they will take the pet to by the lake, taking a walk on the garden and lawns. This small area has several hundred pets about.

6: I am in aid of you resolve to maintain concerning the appliance, the thing industry contact, install the telephone etc. daily life problem.

Hope that you can like our city, our community, our house. The landlord of your enthusiasm: Lijun's soldier.

The community had about 700 apartments, a large garden with walkways but no lawns, a tennis court and a basketball court but no gym, and maybe 100 dogs and a lot of stray cats. The intended meaning of the letter above is

About the community

1: Our community environment is beautiful. There is a lake, a large garden, a tennis court, outdoor exercise equipment, and two playgrounds for the children. Just outside the community's west gate, there is a bank and a convenience store.

2: Nanjing's largest pet hospital is located across the street from the community's west gate.

3: A large supermarket and other shopping is just 10 minutes away. The main shopping district downtown is 20 minutes away. Convenient bus transportation is just outside the west gate.

4: You can have fresh milk delivered each morning to your door, and you can also arrange to have daily or weekly maid service.

5: The community is pet friendly, and you can see many residents walking their dogs during the day.

6: I am available to help you resolve any problems with the appliances, contact the property management company, get the telephone and Internet service installed, or other problems you may encounter concerning the apartment.

I hope that you will like our city, our community, and our apartment. Your enthusiastic landlord, Lijun.

TRAINING SESSIONS AT RGS TECHNOLOGIES

The documents I reviewed ranged from software product descriptions, upgrade guides, user guides, operation manuals, and installation guides to human resources documents, media releases, speeches, policy documents, and regulations. As I reviewed all the different types of documents, I began to see some patterns in the wording and organization, which gave me a wealth of material for the monthly training sessions. For my first training session, I presented issues that I had encountered in the first two documents I had edited. Since I had been working mostly on my own, I tried to make the session as interactive as possible so that I could get to know the 30 or so Chinese translators a little better and so that they could get to know me better and feel comfortable with me.

"Many of you are having difficulty with using articles correctly," I began. "That's understandable, since there are no articles in Chinese."

There were several nods of agreement and a few expressions of relief around the room.

"I can give you some general guidelines—most of which you probably know already—but unfortunately, I can't give you any hard rules. English gets kind of complicated that way."

"You mean, like whether a noun is countable or noncountable?" asked Wang Bing.

"Yes," I replied. "One hard rule I can give you is when to use 'a' and when to use 'an.'"

"Use 'an' before a vowel and 'a' before a consonant," said Wang Bing.

"Almost," I said. "Use 'an' before a vowel *sound* and 'a' before a consonant *sound*. For example, it is 'a uniform' and 'a useful tool' because 'u' in this case sounds like 'yu,' a consonant sound. 'He is an honest man' because the 'h' here is silent and the beginning of 'honest' has the same sound as 'awful.' However, we say 'he is a horrible man' because the 'h' is pronounced here, just like the 'h' in 'hello.'"

"This is where it helps to know how to speak English and hear the sounds," said Lu Hai Tao.

"Exactly," I replied. "Simply reading is not enough. 'H' and 'u' are the two letters that will give you the most problems, so you can also check the pronunciation in the dictionary."

"It could be worse, you know," I continued. "With English, we struggle with whether to use an article. In some languages such as French and Spanish, you have to use articles, but you also have to know whether the noun is considered to be male or female."

"I also want to point out a key difference between American English and U.K. English with regard to collective nouns. In American English, most collective nouns use a singular verb, whereas in U.K. English, they can take a plural verb. We say 'the staff is knowledgeable' whereas the British say 'the staff are knowledgeable.'"

"I often have problems with that," said Zhang Ni, who spoke English with a pronounced British accent. "So if you want to talk about everyone on a team, why can't you use a plural verb?"

"The team is one unit, a single unit. So the team is. However, the team members are. The team works hard, and the team members work hard. If you use the collective noun to modify a countable noun, then you can use a plural verb form."

The translators told Harry that the training was very useful and that they looked forward to the next one. In subsequent training sessions, I covered prepositions (also something that Chinese doesn't use) and nonstandard phrases that I saw in the translators' work. For example, Chinese has a common expression that literally translates to "more and more." An idea may get translated as "Today, the number of Internet users is becoming more and more. This means that the demand for bandwidth is also becoming more and more." The idea can better be translated as "As the number of Internet users increases, so does the demand for bandwidth."

Translators also have trouble expressing time and often misuse words such as "recent," "recently," "since," and "from." For instance, instead of saying "Since its start in 1979, the advertising industry has developed rapidly and grown 40 percent annually for the past 30 years," they would say "Since the advertising industry starts from 1979, it has been developing rapidly with the growth rate of 40 percent every year in recent 30 years."

REQUEST FOR MORE

After the Chinese New Year, Richard asked me whether I knew of any other North American colleagues who would be interested in doing editing for them.

"Let me see what I can do," I told him. I was concerned that we would have difficulty finding people qualified for the job. We didn't just need native English speakers; we needed native English speakers who knew technology and who could write and edit. At that time, the department manager was desperate for people, and he didn't have the same concerns I did. I contacted a few colleagues and friends to see whether they would be interested in working with us.

It is not easy to find qualified technical writers and editors who are already living in a second-tier city in China. In Nanjing, the bulk of foreigners are teachers, students, or nonnative English speakers who work for one of the many multinational companies located here. Chinese companies look to hire foreign talent locally, because they are unwilling to pay high wages for that foreign talent. Although we make much more than our Chinese counterparts, we are paid salaries that are half to one-third of what a technical writer or editor might make in the United States. However, since the cost of living in our city is much lower than in the United States, we make a comfortable living.

By March 2010, we had a team of four Americans and one Canadian. Our team consisted of a retired businessman, a former park ranger, a master's degree student with a bachelor's degree in computer science, a technical writer who was very focused on content but not grammar, and me, a technical editor and writer. All of us have lived in China for two to six years and all of us, except the retired businessman, speak Chinese fairly well. Two people read Chinese quite well, and two of us can read some characters. Four of us have experience teaching English in a university or corporate setting or both. Most of us understand many of the differences in communication styles between Chinese and English. I was responsible for getting three of the team members hired, and one of my colleagues brought in the fifth person.

But not all of us are really editors, per se, so the accuracy and detail of the editing varied. The retired businessman caught the glaring grammar issues but not writing-style issues such as excessive use of passive voice and prepositional phrases and lacked the technical background to catch product description issues. The former park ranger was too colloquial in his edits, and although he was familiar with some technology, he lacks a solid technical background. The master's student has turned out to be an excellent technical editor and is seriously considering technical editing as a career. The technical writer did an excellent job catching technical issues and streamlining verbose and redundant text but did not pay attention to grammatical issues.

ISSUES WITH TRANSLATION

Lu Yan is a close colleague of mine. She is a senior translator whose job is to proofread the translations and check them against the original Chinese for accuracy, ensure that the documents adhere to the company style guide, and make the documents as readable as possible in English before sending them to the North American editors. Lu Yan constantly complains about the poor quality and vagueness of the Chinese source documents.

"Why are the documents so vague?" I asked her.

"If the language is vague, and the readers can't understand what the Chinese says, then no one has to be responsible for carrying out the policy," Lu Yan replied.

"What!" I exclaimed. "You can't be serious! How does anything ever get done?"

"Sometimes, nothing ever happens. The statement is made for the official record, and nothing happens until there is a problem. I think it's really bad, but no one wants to change the meaning."

Chinese speakers can be masters of ambiguity and vagueness with their language when they so choose. This vagueness works well for Chinese society. By stating something vaguely, one avoids the risk of offending anyone and perhaps losing face. Instead of saying "Keep off the Grass," a Chinese sign might say "Please enjoy the lovely tender young grass." (To me, this is an invitation to eat the grass rather than a request to not walk on it.) This is a culture where people often say the same thing several times, each time a little differently, in case the reader didn't understand what was said the first time but doesn't want to ask for fear of losing face—or cause the writer to lose face, since asking questions implies that the writing isn't clear. This is a culture where, more so than in many western societies, people present collections of facts without drawing a conclusion, submitting an opinion, or tying the facts together to make a point. Writers will use metaphors (many of which cannot be easily translated into English) extensively to illustrate what they are trying to communicate, but it is the reader's job to digest the information and determine what conclusions or ideas are to be drawn from the information that has been presented. It is all very "safe," but it does not work for technical documentation or policies and regulations where things need to be specific.

This method of communication is contrary to what westerners perceive as "technical communication." Many non-Chinese find this cultural tendency quite frustrating; Chinese people will simply shrug their shoulders and say, "that's the way we do things in China." We must constantly remind them that the documents are destined for North American, South American, and European audiences, and this is not the way people in those cultures do things.

Consequently, the documents that the translation department receives are often poorly written and extremely ambiguous by both Chinese and non-Chinese standards and are rife with redundant information. The documents often state the obvious while omitting key pieces of information. The translators are faced with trying to wade through the ambiguity in the Chinese writing so they can understand it and then create an accurate, understandable English translation. Often, the translators know that information is missing when they do the translation, but the tight deadlines make it difficult for them to go back to the author to resolve the issues. With the time constraints, too often the end result is a document that has been translated word for word, with no attempt to determine whether the entire piece makes sense or flows logically.

"Why are there so many incorrect verb tenses in this document?" I asked Ding Jia Juan, another of the senior translators with whom I work.

"The style guide says we must use simple present tense," she replied.

"Well, simple present tense doesn't always work," I said. "This sentence says 'The upgrade is failed.' It should use the present perfect tense—'The upgrade has failed'—because the upgrade has already happened. Also, when an operation has been performed successfully, the operation was successful, not *is* successful. The operation has already been performed."

"I got it," she laughed. "Unfortunately, our performance is measured by how well we adhere to the style guide, not how well we follow your advice."

"Ah, so that's why I keep making the same edits and they never get incorporated!" I exclaimed. "I'm afraid the style guide is blatantly wrong in some cases. What can we do to get it fixed?"

"I'm not sure," she replied. "They say Richard wrote most of it."

"Oh," I said. "Just because he is the boss, it doesn't mean he's always right. I need to find time to make a thorough edit of the style guide and make it a priority. We'll never get the errors out of the documents as long as people are adhering to a style guide that has errors. I will even provide cases for why the items are wrong."

By this time, Lu Yan had joined us. "I really hope you can get the errors changed," she said. "Ding Jia Juan and I often know there is a problem with the English, but we are told that we must comply with the style guide."

"Let me see what I can do," I replied.

RGS Technologies has a rather detailed style guide—400 pages—for documents that are produced in English. However, the style guide was written in-house by Chinese writers. No native English speakers, much less any technical editors, have been consulted for the style guide content, so the style guide that everyone must use when translating and editing has many errors in it concerning accepted American English grammar, usage, terminology, formatting, and style. When the North American editors correct documents to have them conform to standard American English, the corrections are often ignored because they conflict with the style guide.

"Do the writers have a style guide for the documents they produce in Chinese?" I asked.

"The company has guidelines for authoring documents in Chinese, and we have templates, but most of the writers don't follow them. There are no processes or procedures in place that allow us to enforce the writers' adherence to the guidelines," replied Lu Yan.

"Does anyone edit the Chinese documents before you get them for translation?"

"No. The company thinks that because we are Chinese, everyone can write in Chinese," Lu Yan rolled her eyes. "The lack of consistency in the Chinese documents makes the translators' jobs more labor intensive. And it also makes the proofreaders' jobs harder because the inconsistencies in Chinese create even more inconsistencies in the English translations."

"In the United States, we know that not everyone can write. That's why people like me always have a job," I chuckled. "We have similar issues in the United States with technical people. They are usually better at providing specific information, but often it is poorly stated and poorly organized. I will often interview them and then write the documentation myself, which makes most of them happy because they don't enjoy writing. Although they are subject matter experts, most of them recognize that they are not experts in the English language, and they consider the writing and presentation to be the technical writer's job. The technical people are responsible for the accuracy of the content, but they do not have the final say on how the content is written, as long as it is technically accurate. They know that they don't 'own' the documents—the company for whom they work does."

"Here, the Chinese writers have the final say about our translation. Most of the writers can read and understand written English, and they will reject a translation if they feel that it is not true to the original Chinese. As long as this continues, we will have some problems," said Lu Yan.

"Do you think they will ever let the two of us train the engineers and software developers in technical writing so they can learn how to think about presenting information logically and concisely? I think we would do well as a team."

"I don't know," said Lu Yan. "But I would really like the opportunity to work with you on that."

Lu Yan has a good reputation as a trainer, and I felt that we would be able to work well together. Although my Chinese is fairly good, I am by no means fluent, and I knew I could trust her to field questions effectively. We would be a "dynamic duo." Unfortunately, we are still waiting for management to seize the opportunity that we have presented to them.

CHANGING TIDES

In April 2011, RGS Technologies decided that they wanted to have foreign managers (team leaders, really) manage the documentation projects for each of its four locations (Shenzhen, Shanghai, Nanjing, and Beijing). I was chosen to manage the Nanjing office. Those of us who were chosen were offered a miniscule raise and promised many things. Then, while I was on vacation, the company decided that they didn't want foreign managers and took away our miniscule raise. They never said a word to me. I got wind of what was happening because the Beijing manager called me. When I returned from vacation, I was simply told that there were no tasks for me. The day after I returned, the translation company that had been handling my employment since July 2010 called me and asked me to come into their offices. They offered me a position at the rate I had been paid before the demotion and I accepted. Apparently, one of the editors at RGS Technologies' Shenzhen office complained about me and I was terminated without my knowledge. You see, while I was on vacation, I was doing editing work remotely. I suspect I may have made some comments about a speech that upset someone. Many times this speaker would say things that I didn't agree with, but I always held my tongue and corrected only the grammar. However, three times I corrected factual errors. If I had to do it over again, I would have done the same thing. I am still doing editing for RGS Technologies, but I am now working offsite.

Unfortunately, I was unable to review and edit the corporate style guide. The management made a few corrections and published a new version, but it still contains numerous errors. My "spare-time" project was to review it thoroughly. Since I seemed to be the only one interested in pursuing the erroneous style guide, I suspect that it will go untouched.

RGS Technologies has, to its credit, recognized there is a problem with its documents not sounding "native." However, as long as the Chinese authors (who have no training in technical writing) are the "owners" of the documents that they write and have final approval of the translated documents, the company will never be able to produce documents that are ready for global audiences.

There is also the issue of cultural understanding. My story has focused mainly on text-level translation because until companies can get past these issues they cannot successfully localize their products and documentation for audiences from different cultures. But Chinese companies also need to understand that writing documents for global audiences is not simply a matter of translation. It is always easier to translate from one's second (or third, fourth …) language to one's native language because we know the cultural context of our native language best and are most comfortable with it. Depending on how much culture we learned when acquiring our second (or third, fourth …) language, we may or may not know enough cultural context for the nonnative language. Although most Chinese translators in Nanjing (and most of China) have spent years studying English, they have not had much opportunity to speak English or using it to actually communicate with native speakers. Although U.S. television shows such as *Friends, Desperate Housewives, Gossip Girls,* and *The Big Bang Theory* and U.S. movies have become extremely popular in China, they represent popular culture and make-believe realities, not the real-world business environment or people's lives in the United States.

Chinese companies and the translation industry in China need to be able to hire people who are not only fluent communicating in another language but who have spent time abroad or are familiar with cultures besides Chinese culture. Without knowledge of the other language's "culture," successful translation and localization is not likely to occur easily.

RECOMMENDED READINGS

1. Melanie's experience is that Chinese documentation, because of its language and cultural influence, is conceived and composed in a way that is very different from English documentation. Other practitioners and scholars may not necessarily agree with this statement or have different perspectives in comparing Chinese and English documentation. The following readings deal with this topic:

- D. Ding, "The emergence of technical communication in China—Yi Jing (I Ching): The budding of a tradition," *Journal of Business and Technical Communication,* vol. 17, no. 3, pp. 319–345, 2003.

- D. Ding, "An indirect style in business communication," *Journal of Business and Technical Communication,* vol. 20, no. 1, pp. 87–100, 2006.

- H. Yu, "Putting China's technical communication into historical context: A look at the Chinese culinary instruction genre," *Technical Communication,* vol. 56, no. 2, pp. 99–110, 2009.

2. The author briefly discussed the concept of *guanxi* and how it is essential for doing business in China. The following sources discuss this concept in more detail:

- K. St.Amant, "Considering China: A perspective for technical communicators," *Technical Communication,* vol. 48, no. 4, pp. 385–388, 2001.

- G. Bissky, *Wearing Chinese Glasses: How (Not) to Go Broke in Chinese Asia.* Victoria, BC, Canada: Trafford Publishing, 2007.

- D. Y. Lee and P. L. Dawes, "Guanxi, trust, and long-term orientation in Chinese business markets," *Journal of International Marketing,* vol. 13, no. 2, pp. 28–56, 2005.

- C. Ho and K. A. Redfern, "Consideration of the role of guanxi in the ethical judgments of Chinese managers," *Journal of Business Ethics,* vol. 96, no. 2, pp. 207–221, 2010.

3. This story briefly mentioned the importance of cultural understanding in translation. The following sources provide more information on how language and communication are intrinsically cultural:

- K. St.Amant, "When culture and rhetoric contrast: Examining English as the international language of technical communication," *IEEE Transactions on Professional Communication,* vol. 42, no. 4, pp. 297–300, 1999.

- A. Wierzbicka, *English: Meaning and Culture.* New York: Oxford University Press, 2006.

4. This story commented on the use of automated translation. The following resources provide more information on this topic:

- S. Hurst, "Automated translation for technical documentation—Can it deliver what it promises?" *tcworld,* November 2009. Available: http://www.tcworld.info/tcworld/content-strategies/article/automated-translation-for-technical-documentation-can-it-deliver-what-it-promises/, accessed October 24, 2002.

- U. Muegge, "Controlled language—Does my company need it?" *tcworld,* April 2009. Available: http://www.tcworld.info/tcworld/content-strategies/article/controlled-language-does-my-company-need-it/, accessed October 24, 2002.

- D. Qian, "Prospects of machine translation in the Chinese context," *Meta: Translators' Journal,* vol. 50, no. 4, 2005. Available: http://www.

erudit.org/revue/meta/2005/v50/n4/019854ar.pdf, accessed October 24, 2002.

DISCUSSION QUESTIONS

1. This chapter discussed the nonstandard English usages among Chinese speakers. Now, consider these usages in light of the concept of world Englishes, which holds that English goes beyond British Standard English (BrSE) or American Standard English (AmSE). Rather, different countries and people appropriate the English language for their own use, hence the idea of world Englishes. Indian English, for instance, has become recognized as a distinct type of English. It mixes American and British English with vernacular words, syntax, and direct translations. From this perspective, can we see Chinese English as a form of world English? Is it an identifiable type of English appropriated by the Chinese people? Why or why not?

2. What is the promise of using automated translation tools to translate technical and engineering documentation? Conduct research on these tools to understand their potentials and limitations.

3. At RGS Technologies, Chinese engineers and software developers "own" the documents they author. They also enjoy a higher status than translators and have the right to approve or disapprove their translation. Does the same situation exist elsewhere where technical communication is a more established discipline, such as in the United States and Europe?

4. What are some of the linguistic, rhetorical, cultural, and social factors that make it difficult for Chinese documents to be ready for global audiences? What might it take for those documents to be ready for global audiences, including and beyond what the author suggested?

5. What are some important factors professionals, Chinese or otherwise, need to consider if they want to work in the technical communication field in China? What are some important qualities, experiences, or attitudes one has to have in order to succeed?

CHAPTER 11

SUBJECT MATTER EXPERT MEETS TECHNICAL COMMUNICATOR: STORIES OF MESTIZA CONSCIOUSNESS IN THE AUTOMOTIVE INDUSTRY

Angela M. Haas

Angela M. Haas worked as a technical communicator in the automotive industry for over 10 years and is currently an assistant professor of professional writing and rhetorics at Illinois State University. Her research interests include cultural rhetorics, digital and visual rhetorics, decolonial theories and methodologies, and transnational feminist, indigenous, and technical com-munication studies. Among other places, Haas's work has been published in dozens of procedural manuals, proposals, and audit reports as well as Journal of Busi-ness and Technical Communication, Computers and Composition, Pedagogy, *and* Studies in American Indian Literatures.

CHAPTER SYNOPSIS

This story is inspired by Gloria Anzaldúa's *mestiza consciousness*, a world-view that challenges colonial stereotypes that portray mixed-race people as tragically "caught in between two worlds." Celebrating *mestiza conscious-ness*, this story presents multiple promises of occupying hybrid identities. In the story, we get to know technical communicator/QS consultant Angela, sea-soned factory worker Hermoso, and their colleagues at a tier 2 automotive parts supplier on the south side of Toledo, Ohio. These people worked togeth-er to help their plant meet the automotive industry's certification requirement. By telling their lives, this story attempts to capitalize on the agency of hybrid

Negotiating Cultural Encounters. Edited by Han Yu and Gerald Savage
Copyright © 2013 The Institute of Electrical and Electronics Engineers, Inc.

identities and negotiate the distance between several knowledge systems and discourse communities. The story highlights some of the tactical inventions and interventions that happened in the liminal spaces between subject matter experts and technical communicators and some of the ways in which hybridity may have influenced the work and relationships in these spaces.

> *The struggle is inner: Chicano, indio, American Indian, mojado, mexicano, immigrant Latino, Anglo in power, working class Anglo, Black, Asian—our psyches resemble the bordertowns and are populated by the same people. The struggle has always been inner, and is played out in outer terrains. Awareness of our situation must come before inner changes, which in turn come before changes in society. Nothing happens in the "real" world unless it first happens in the images in our heads.*
>
> En unas pocas centurias, *the future will belong to the mestiza. Because the future depends on the breaking down of paradigms, it depends on the straddling of two or more cultures. By creating a new mythos—that is, a change in the way we perceive reality, the way we see ourselves, and the ways we behave—*la mestiza *creates a new consciousness.*
>
> —Gloria Anzaldúa

After working for nearly two years as a technical communicator at a production facility (factory) for a major U.S.-based rubber company with tens of thousands of employees worldwide, I accepted a position with a young, small startup company with 100+ employees that manufactured glass hardware and seat track assemblies for the automotive industry. Although both factories were located in the borderlands and shadows of the Motor City (Detroit, MI) and Glass City (Toledo, OH), both ultimately served the Big Three (Ford, General Motors, and Chrysler), and both hired me to manage their documentation systems, each required a unique *mestiza consciousness* in order to succeed. This story comes from the second job.

MEETING HERMOSO

I had been following Hermoso around the plant for about a week, scribbling reams of notes on his day-to-day work practices on pads of paper secured to a wooden clipboard. (All character names are pseudonyms.) I observed, listened, asked questions, and recorded what I learned.

"You're not like the other office employees," Hermoso claims loudly over the cacophony of whirrs, clangs, rattles, hums, slaps, swooshes, shuffles, beeps, and thumps emanating from the interface between workers and windows, brackets, assembly machines, glazing machines, curing chambers, testing instruments, forklifts, semi-trucks, and other technologies on the factory floor. He continues talking during his routine, preproduction setup of the XJ fixed rear vent window for Jeep Cherokees. "They usually just stay up in the front office at their computers ... until a problem happens. Why do you need to talk to us operators out here on the shop floor when your office is next to the process engineer who drew up the plans?" he questions, sarcastically.

Hermoso positions an empty finished goods container on the conveyor next to his machine and the raw glass container on the cart facing him. He grabs the closest piece of raw glass and secures it on the machine. At first, a couple of his colleagues operating the other XJ machines in this work area turn to look on, but they soon lose interest in my presence and Hermoso's instructional discourse and return to their own setups and assemblies. I carefully watch Hermoso and feverishly jot down notes as he walks me through the assembly process, from start to finish—including verifying the parts required to complete the assembly, assembling the finished product, inspecting the product for quality, and packing a full finished goods container—and we ask questions of and answer one another along the way.

Hermoso efficiently tends to some housekeeping around his XJ work area and relocates to an adjacent workstation designed for assembling SN-95 Mustang quarter windows. One of the earliest employees of this small tier 2 automotive parts supplier on the south side of Toledo, Ohio, Hermoso knows the ins and outs of completing every new and existing job contracted with this growing company. He was hired not long after the company was established to do business with and for Ford in the late 1980s. Less than a decade later, he had had extensive experience working every machine required to meet the existing contracts with General Motors, Chrysler, and Ford, or the Big Three, as many in the automotive industry collectively dub the three companies. Therefore, it was immediately clear to me (and Garrett, the manufacturing manager who asked Hermoso to mentor me) that Hermoso's experiential knowledge of the everyday operations of the company was going to be critical to my success in my new position as the company's first technical communicator.

MY WORK IN QS-9000 CERTIFICATION

I was hired because the Big Three had recently required their suppliers to become QS-9000 certified (superseded in 2006 by ISO-9001 and TS16949), a renewable certificate of management and manufacturing quality used in the automotive industry. This certificate was based upon the ISO-9000 quality system standards (published by the International Organization for Standardization). The QS-9000 quality system manual (published by the Automotive Industry Action Group) was long, boring, and full of industry-, customer-, and standard-specific jargon. The first section of the QS-9000 quality system manual adapted 20 required elements of the ISO-9000 standards to the automotive industry, and the second section included requirements for meeting customer-specific expectations communicated by the Big Three and other original equipment manufacturers (OEMs). As with my first technical communication position, I was hired for this job because no office employees wanted to read the manual and figure out how to implement it. I had nearly two years of prior technical writing and QS-9000 experience with a major rubber and tire company (a tier 1 supplier to the Big Three). When I showed up to the interview with my well-annotated copy of the manual, they offered me the job on the spot. I was charged with coordinating all of the necessary documentation to achieve the certification required to continue doing business with the Big Three (and the Big Three's tier 1, tier 2, and tier 3 suppliers).

Despite my prior technical communication and automotive industry-specific experience, this job was different. I was immediately thrust into a unique, complex network of power relations and negotiations—between factory management and the Big Three, between the office and floor personnel, between human resources (HR) and the union representatives (union reps), and between the union reps and other union members.

The workforce at my former place of employment had been overwhelmingly comprised of Euro-American workers. Here, approximately 80% of the factory personnel primarily self-identified—reflecting histories of U.S. colonialism, imperialism, and ethnoracism—as one or more of the following complex and sometimes overlapping ethnic and political identities: Mexican, Mexican American, Latino/a, and/or Chicano/a. A few also used "Hispanic." If helpful, a brief cultural–political lesson follows: Latino/a includes those with ancestry from or born in Latin America, including Mexico; Chicano/a signifies a political affiliation within Latino/a culture concerned with affirming the self-determination of Mexicans and Mexican Americans who challenge the legacies related to the United States having stolen Mexican land and renaming geographies and people; Hispanic is often used to refer to collective peoples and populations who are or have been under Spanish rule. Moreover, about 75 percent of these employees were women.

Given all of this, although I was literate in the rhetoric of QS-9000 standards and in ways of implementing the requirements as part of a quality documentation system that supports the management system, I did not necessarily have the cultural, informational, linguistic, or technological/mechanical literacies required to ease into the job responsibilities and workplace cultures. Thus, I needed Hermoso and his coworkers to share their subject matter expertise regarding the machines, processes, and workplace cultures in order for me to do my job well.

I followed Hermoso in hopes of learning how to work and, subsequently, document the work instructions for operating the dozen or so active machines on the shop floor, each one of them responsible for a different window assembly, glazing process, or curing process. In doing so, I learned much more than this. I learned how to navigate my way around the well-cleaned transportation paths around the well-used machines—some of them coated with fresh adhesive from the assembly process or grease from routine maintenance. I learned in which direction to dodge the fast-moving forklifts in specific areas of the plant. I learned how to decode the colored tape adhered to the concrete floor and which types of areas they were demarking (e.g., for travel or quarantine). I learned that Hermoso knew every worker on the shop floor and was multilingual, calling out to each of them in Spanish, Spanglish, English, or Nahuatl (also known as Aztecan, a language indigenous to Mexico) but rarely stopping to chat. When on the clock, Hermoso walked with purpose and zest, and it was sometimes difficult for me to keep up with him as he zipped around the warehouse in his signature ribbed, bright white "wife beater" tank and well-worn low-rider jeans. However, when on break, he enjoyed laid-back, lively conversations with his friends and family who happened to be coworkers.

I also learned that Hermoso, a late-40-something, hard-working Mexican American man with a long history of working for all Euro-American male managers, had plenty of reasons to distrust management and Euro-America. But for some reason—

that he did not articulate and I may never know—he trusted me. Perhaps because I am concerned with how race, class, gender, and more can influence relationships and people in workplaces. Perhaps because I am a younger, working-class, mixed-race (primarily German American and Eastern Cherokee) woman. Or perhaps he just could sense that I was trustworthy—because I am; I trusted him, too.

Because of this mutual trust, I learned that he had work-arounds for almost every machine, tactical moves that assisted him in meeting his assembly quota for each machine in a more efficient manner than first imagined by the process engineers. Although he trained me in how to operate each machine and assemble the parts required of the job so that I could write the procedures and corresponding work instructions per the engineering specifications, off-the-record, he also showed me his work-arounds. Based upon Hermoso's prior experiences, I agreed to leave these work-arounds out of the documentation to be published in the company standard operating procedures (SOP) manual. Why? The most recent occasion of Hermoso sharing his tactical knowledge with management resulted in management thanking Hermoso and making him employee of the month, with his name permanently etched on the plaque prominently displayed in the front reception area. However, management also changed the engineering specification and subsequently raised the daily quota of assemblies required of workers on that production machine, ultimately ignoring variances in able-bodiedness across the workforce and limiting their potential to earn "overquota" bonuses.

VENDING MACHINE LUNCH AND TAMALES

After my first week of typing up my notes, printing them, and taking them out to Hermoso to ask for feedback on the accuracy of my representations, I attempted to write the first set of documentation required for each element of the quality system. Four levels of documentation were required: level 1, policies; level 2, procedures; level 3, work instructions; and level 4, related forms. I easily mocked up a design for the page layout and a numbering system for these four levels of documentation, but I was daunted by the writing tasks ahead. The learning curve was steep, as each machine had its own interface, safety features, parts required for assembly, and more. Given the climb ahead, nourishment was needed. Time to take my lunch break.

I didn't bring a lunch nor do I feel like taking the time to drive beyond the South Side for something other than commercial fast food, so I decide to piece together the best meal possible from the available options in the vending machines: a diet coke, a bag of pretzels, and a pack of low-fat vanilla crème cookies. I walk past the conference room where many of the women in the front office eat, open the door to the shop floor, turn to my left, and head toward the vending area. There sits Hermoso unusually alone at one of the picnic tables about 8 feet in front of the vending machines. "Hey Hermoso," I call over as I near the area, just as he takes a big bite of pork tamale. He shoots me his friendly smile, nods toward me, and continues chewing. I feel his eyes on me as I feed the vending machines, make my selections, and snake my arm into the narrow dispensing receptacles.

I turn around from the vending machines, and as I walk toward the picnic table to join him, Hermoso fires an arsenal of questions, as if he'd been saving them up for me. "What's up with this quality policy we're all supposed to memorize as part of this QS-9000 stuff? Do we really have to know every word of it, by when do we need to know it, and who's going to be asking us, what happens if we don't get it right, and why does this thing matter anyway? I mean, we have a good record of producing good-quality products. How is a policy supposed to change that one way or another?" It becomes clear that Hermoso learns new concepts by carefully thinking on them for a good while before asking a series of direct questions that skillfully cut to the heart of matters.

Sitting down across from him, I attempt to fully answer his questions by first breaking down the purpose of the standards and our adherence to them. "Basically, the Big Three want us to say what we do and do what we say. So I am just supposed to be documenting what it is that we value here and how we carry out our daily processes with these values in mind. You follow?"

Hermoso nods, and I continue. "To make sure we catch any inconsistencies before the external certification audit, I will audit all the documentation we are drafting to determine whether what we do and say we do are really aligned and to ensure that what we're doing meets the QS-9000 standards."

"What if it doesn't match up?" he questions.

"Well, the first one is an easier fix—if our actions don't match what we've documented, then I'll just need to revise those documents to better reflect what we're doing—but if we're not meeting the standards, then we'll be required to brainstorm ways of meeting those standards in ways that make sense for our company. And getting back to your question about the quality policy, no, you don't have to memorize it. That's not the point. The Big Three want us to have a quality policy that the entire company values, so that is what the management team tried to draft: a three-sentence policy that encompasses our company's existing commitment to quality manufacturing. So the point of the policy is not to memorize it; the policy is just supposed to articulate the company values that already inform your everyday practice."

"So what exactly will we be audited on, then?"

I respond, "Basically, you'll be asked to explain how the essence of the quality policy impacts what you do on the job every day."

"So what you're saying is that we can boil those three sentences down to the essence of what they mean to us?" Hermoso queries.

"Exactly. As long as you make sure it still contains the same ingredients after you boil it down," I respond with a smirk.

"That's cool. I can do that." He cautiously proceeds, "And as long as we do the jobs per the instructions when we're being audited, we're cool, right?"

"That's correct."

We sit there in silence for a few minutes as we eat our lunches. Without looking up, he comments, "That crap will kill ya, ya know. All those sugars and artificial ingredients." He looks up. "You see my sister over there? The one in the long grey t-shirt." His hand gestures and his head jerks toward a heavy-set woman assembling Jeep window components, her arms and legs working away as she repeatedly pivots at the waist at her assembly machine, flanked on either side by two other Latinas at two other workstations. They work almost in unison.

I nod, and he continues, "She's been eating those low-fat cookies you're eating there everyday for the last three years or so. You think she's lost any weight? The good news is that you know the ones you're eating are fresh since my sister forces a high turnover in that vending machine." He grins. "Well, as fresh as fake food can be." His grin widens.

I counter, "But your lunch may be just as likely to take you out, but the difference is that I can't call tamales crap or fake—that is, if they're good."

"What do you mean?" he inquires, still smiling.

"I mean, if that there is a good tamale, then it is made with lard and cheap cuts of pork or beef. And those cheap cuts of meats typically have a lot more fat on them. And that's probably not so healthy either. But then, again, if it's made that way, and with just the right balance of masa and meat, then that's probably about the best lunch I could ever imagine. Nothing better than eating the perfect tamale."

He looks at me quizzically as he eats the last of his second and last tamale. "How do you know so much about tamales?"

"My two best friends from growing up were Mexican. Well, I guess they still are. Well, one is. Umm, they both are; it's just that one has since passed. But anyway, sorry ... that was awkward." He gives me a reassuring look, and I press on, "To answer your question, I basically spent about half of my preteen years in either Angie Mendieta or Christina Baldonado's house, and their mothers fixed us the absolute best homemade tamales, chorizo and eggs, tortillas, and more that I have ever eaten."

"Well, my aunt and my sister ... another sister, not that one," he gestures again, "hand rolled the tamales I just ate. And I assure you, my friend, that those are in fact the absolute best homemade tamales. Oh, and that reminds me, speaking of that other sister, she works at Chrysler, and she doesn't have to do this QS-9000 crap. I guess since Chrysler's one of the companies holding the purse strings, they get to ask others to do as they say, but not as they do."

"Now that you mention it, I don't really know what the Big Three are doing regarding these standards that they're asking us to comply with. I think I've just blindly assumed that these standards are standards that deserve to be met and that they, of course, wouldn't ask us to follow them if they aren't following them. Maybe they are implementing them, too, but just aren't communicating what they're doing to all of their employees. I cannot really say one way or another, to be honest. You would likely know better than me since I don't know anyone who works at any of the Big Three around here—or Jeep, for that matter. My family's only worked in factories that supply the Big Three—or in other industries, like the textile mills in the Carolinas."

"Is that right?" he asks with surprise. "You must not be from around here then because I thought you couldn't throw a rock without hitting somebody who works at one of those plants. There's so much inbreeding up here in these factories. I mean, look around. Between my sister, my sister-in-law, and a couple cousins, there's five of us that work here. And some of my friends, too. Everyone working here has a similar story. I mean, look at our union reps: they are a father and son duo." He gestures and jerks his head in the direction of two, weathered blond guys of similar build and stance standing in front of a machine, seemingly in conflicting dialogue about something.

"Well, I'm from Bowling Green. I'm pretty sure my parents know people who work at the Big Three. I just don't personally. But it's similar at my mom and dad's plants. In fact, over the years, my mom has gotten a couple of her sisters and my older brother a job at the factory I left for this job here."

"You're talking office jobs, right?" He asks.

"Well, it took my mom 20 years to work her way up to the front office, after working on the factory floor as a forklift driver, machine operator, quality auditor, and more. And my dad worked his way up from the mailroom to a shipping clerk at that major glass factory here, and now he is the shipping manager. So his offices moved over time from a basement office to a 'back office' at the far back corner of the plant and finally to the 'front office.' But they both were the first in their families to eventually be considered white collar workers, as I come from a long line of factory and mill workers and farm hands."

Hermoso downs the last drink of his coffee as I crumple up my lunch wrappers. He responds, "Well, who knew we would have so much in common?!"

"Who knew, indeed! And I've enjoyed getting to know all these things that we have in common," I add as we both stand and clean up around our spots at the table. "Well, I guess it's back to work. Have a nice day, Hermoso."

"It already is," he says matter-of-factly, as though it is a pat response. "You, too."

We both walk toward our workstations, his out on the floor toward the assembly machine he was operating that afternoon and mine a cubicle in the front office. Almost to the front office door, I think I hear my name being called over the factory noise. Turning around to verify, I hear a whistle. Here comes Hermoso extracting two pinkies from his mouth as he rushes toward me. I walk toward him, and his pace slows. "Sorry to whistle at you," he offers, nearly out of breath. "I was just trying to catch you before you got up front."

"That's okay," I assure him. "I understand. Plus, I'm intrigued. I have never seen anyone but my father whistle like that. Using your pinkies like that."

"You don't get out much do you, Angela?" he jabs with another sheepish grin. "So I got back to my machine, and it came to me. My take on the quality policy. And I want to know what you think, like whether it would meet the standards."

"Okay, shoot."

"*Zero defects, continuous improvement,*" he asserts. "So what do you think?"

"Sounds good to me. It touches on the essence of the policy and works for several stakeholders."

"That's what I was thinking," he interrupts excitedly. "I mean, I know we want our customers to think we're committed to them, and we do that by always shipping good products. And since Matt was hired as the continuous improvement manager, '*continuous improvement*' is being thrown around everywhere. It sounds good to customers that we care enough to talk about improving what we do, right?"

With a playful smile, I respond, "So are you suggesting that we're just all talk, Hermoso?"

Although the question is a rhetorical one, Hermoso answers anyway. "No, you know from our previous conversations that continuous improvement is more than a word to me, and most of us, really. I think it's the foundation for the entire quality policy, but who wants to memorize all that? Why not just say that we value continuous improvement, and then each one of us is responsible for figuring out what it means to our job. I work to continuously improve what I do not just for our customers, but for me and my family. I want to be a good role model to my kids. I take pride in and want to be the best at what I do. And I want to continue to provide for them."

"And it looks like you're doing a good job of all that, given that plaque on the wall in the entryway to the front office," I acknowledge.

"It's probably the eighth time or so I've gotten it. I've been nominated even more than that."

"I'm not surprised, Hermoso."

"Gracias."

Realizing that I lost track of his series of concerns, I steer the conversation back to one of Hermoso's unanswered questions. "Getting back to your idea for your own quality policy mantra—*zero defects, continuous improvement*—my first reaction is that it meets the requirements. Even more so, it seems to meet your quality values. So it's kind of like your very own personal quality policy. But your personalized policy may also resonate with many of your co-workers. Both up front and back here. Let me think on this some more, and I'll get back to you tomorrow. Muchas gracias, Hermoso."

"De nada, Angela."

MATT, THE CONTINUOUS IMPROVEMENT MANAGER

Energized by Hermoso's enthusiasm, I return to my shared office. Matt is back from his meeting with Tim, the engineering manager, and sits at his desk in our office. It was Matt and Tim who hired me to be the documentation specialist, QS or ISO-9000 consultant, or whatever else they decided to call me on any given day. It really depended upon the context. If they wanted to ask a question about the standards, I was the QS or ISO expert. If they wanted me to proofread a memo for confusion or typos, I became the tech writer. I didn't have a job description, so that meant I was going to have to write one for myself. After all, QS-9000 required HR to document and maintain job descriptions for all quality-related personnel. But the writing and revising of job descriptions wouldn't come for months, as they were classified as level-4 documentation under the training element in the QS-9000 standards—the 18th element out of 20, in fact. I was still working on the first two levels of documentation required for the first two required elements of the QS-9000 standards. The first element was management responsibility, and the first responsibility listed was the necessity of a documented theory and practice of quality, specifically a quality policy, so it was perfect timing for Hermoso to bring me his personalized quality policy.

"Welcome back, Matt. How'd the meeting go?" I plop down at my makeshift workspace. Still waiting on my "real" office furniture, I made do with an unstable, "wood" veneer table with metal legs and a well-used office chair that no longer allowed the user to raise or recline the seat. My workspace was located just inside the flimsy modular office door attached to modular walls that collectively repurposed about a third of a used warehouse into office space. Interestingly, though, the walls for all the offices other than those of the plant manager, HR manager, and accounting manager stopped about a foot short of the ceiling. Thus, it was difficult to secure privacy when speaking with others in our offices, whether by phone or face to face.

"Oh, great, Angela. Thanks for asking. Everything copacetic?" Matt asks from his workspace in the opposite corner of the office.

"Copacetic?" I reply.

"Oh, did I not use that right? I just learned it this morning on my word-a-day calendar. I figured I'd try it out on the English teacher. Probably not a good idea for my first attempt, huh?"

"Oh, no, it's me; not you. I just wasn't anticipating the question."

Leaning back in his plush, adjustable executive chair, he clarifies, "I was just wondering if everything was going okay with the documentation since you weren't in the office."

"Actually, I took an early lunch and had a fascinating talk with Hermoso that may have an impact on our quality documentation system. I think that we should make our work with QS-9000 transparent to all employees as we're doing it. That way we can get feedback along the way. And I've already learned so much from Hermoso. For example...."

I fill Matt in on the QS-9000 related conversations with Hermoso, and Matt soon shares my excitement over how Hermoso was processing what he was learning about QS-9000 and how much we could learn from his willingness to share this with me. Matt's face and eyes lit up in a way that made it evident to those around him every time he brainstormed what he thought was a bright idea. Whatever he had in mind, Matt was on the job with boyish enthusiasm. "I really didn't want to have to get back to reading that boring QS-9000 requirements manual, anyway. So I'm off to make things happen, and you can have the computer for the rest of the afternoon. Oh, and I cleared out the bottom drawer of the filing cabinet for your use until your furniture gets delivered."

"So does that mean my furniture has been ordered?" I query.

"I guess I better take care of that, too, then." Matt claps his hands as if to pep himself up for his upcoming tasks, turns, and rushes out the door and down the hall to meet with Jay, the plant manager, and Jerry, the HR manager. I later learned that his mission focused on talking about ways for communicating QS-9000 across the organizational chart in order to secure plantwide input and buy-in—and one of those strategies included promoting Hermoso's personalization of the quality policy as a plantwide policy for answering external auditors when they ask about our quality policy. He also finally made the time to order my own computer, grown-up desk, filing cabinet, and so on.

OPEN MEETING

One of the first results of Matt's meeting came less than a week later when the trio of managers held an open meeting with the plant employees. They put on their safety glasses, walked out into the middle of the plant floor in their steel toe boots, and asked all the employees to take a break and gather around. The office employees came out, too, but stayed in the lunch area, where no safety gear is required in spaces contained by blue taped floor. The office employees comprised one-fourth of the circle and stood opposite from the managers, while the floor employees assembled to complete the remainder of the circle. With several layers of people forming rings round the managers, the scene initially reminded me of politicians delivering a stump speech and working the crowd with their cronies. But after just a couple minutes, the scene could be read quite differently. They initially provided a brief overview of QS-9000, the specific standards that most directly applied to the employees in manufacturing, quality assurance, and shipping, and why we had to get on board (read: stay in business and hopefully become more competitive, essentially trickle-down rhetoric from the Big Three). Then the meeting took a turn.

"You see," Jay, the plant manager, explains, "since we don't have a choice as to whether we *want* to get QS-9000 certified, we do want to build and support an environment where it is possible to obtain our certification in ways that sustain the values of our company and the people who work here." Jay then holds Hermoso up as a model of how that can happen—through conversations with the new QS-9000 coordinator (they pointed in my direction and reintroduced me in case some didn't have the chance to meet me yet) and Matt, the two primarily responsible for making our certification happen.

Hermoso's colleagues cheer and clap for him. They ask for other ways of sharing ideas and providing feedback for making the certification process work for us. Employees offer a range of ideas, from wooden comment boxes affixed to the walls for anonymous suggestions to a formal procedure, instructions, and forms for providing documented and attributed suggestions. The father and son union reps ask how employees might be rewarded for helping the company in ways that go beyond the employee-of-the-month picture on the entryway wall that so many of them don't even get to see since most enter the building through a different door. After the HR manager, Jerry, reminds them of the $100 cash bonus (adding that had one of them been awarded the employee of the month honor, they would have known this), he says he is willing to consider other forms of compensation.

The meeting lasts about an hour with many Q&A exchanges, and Jay ends with a caveat that employee suggestions will be taken seriously. Further, Jerry announces that we should expect the training on documentation to transpire in waves as sets of documentation are finalized. He explains that the training will provide the evidence we need in our first audit to demonstrate that we all have been trained on the parts of the quality management (and now documentation) system that apply most to our jobs.

The next few weeks consisted of a flurry of QS activity. My office furniture was delivered, at roughly the same time a critical component to phase 1 of Matt's grand QS-9000 plan arrived: laminated "personalized" quality policy cards. The size of a typical business card, one side provided the full version of the quality policy: all three sentences; 100 words in 8-point Times New Roman with narrow tracking crammed into the small space. The other side of the card prominently displays in 24-point Arial, italicized, the company logo and *Zero Defects, Continuous Improve-*

ment. The moment Matt opened the box of cards, he squealed with delight and immediately shot down the hall to show them off to the plant and HR managers. Afterward, he spent half a day distributing them among the floor personnel and chatting everyone up. In the process, he talked to them about quality policy, what it meant, and how to respond to the QS-9000 auditors if asked about the quality policy. He recommended that each employee keep a copy of the card on their "persons," preferably in the back pocket, so they could respond to the questions "what is the quality policy, and what does it mean for your job?" at any time. Finally, Matt asked each employee to sign a training record that documented their understanding of and responsibility for knowing the quality policy. This record could then serve as evidence to auditors that we were meeting both quality policy and training requirements.

Matt also successfully stretched out over an entire week what should have taken him no more than a day or two to complete, affixing two comment boxes on walls in the plant: one next to the door to the front office; the other, along with a duplicate copy of the framed Employee of the Month picture, just inside the plant entrance. (The permanent plaque with specific names engraved on individual copper plates remained at the office entrance across from the receptionist's desk.)

JUST-IN-TIME (JIT) DOCUMENTATION

While Matt was out "making things happen" in the plant, I was frantically writing in the office on our shared Dell desktop. There were at least five different documents open and running in Word. None of them complete, much less ready for feedback, review, or approval.

I had originally mapped out a clear linear plan for completing the documentation required to support the quality management system at the end of my first week on the job. I would first complete the level 1 and 2 documentation (policies and procedures) for Section I of the QS-9000 manual, addressing each of the 20 elements in order—much like I did at my former job. And then I would go back and fill in the level 3 and 4 documentation (work instructions and corresponding forms) for each element. Finally, I would address the customer-specific requirements in separate documentation, if not already addressed in some other way. The first month went mostly according to plan. The scope of the first four elements of QS-9000 (management responsibility, quality system, contract review, and design control) primarily concerned documentation related to the management of the quality system. The management review team was relatively small, consisting of the managers of the plant, HR, accounting, materials control, continuous improvement, manufacturing, engineering, and quality assurance. Therefore, it was fairly easy to write our company's quality policies and procedures for meeting these requirements and to get them reviewed and approved quickly.

But as I began to carry out my plan the following month, I was constantly being pulled in one direction and then in another, simultaneously working on numerous documents at various levels within the documentation system and for several different nonsequential elements. This was due, in part, to the increased involvement of other employee stakeholders in level 3 and 4 documentation for the management-focused elements as well as the manufacturing-focused elements of the standard. Further, before I could finish the documentation for one procedure or set of instructions, we would get a new job in the plant, and instead of retrofitting our documentation, we wanted to establish it properly within our QS-9000 system from the start. I was not used to this. My former employer had a long-standing list of jobs for long-term customers, whereas my current employer was always looking to grow and would take jobs whenever and however possible. Always looking for opportunities that might result in leasing the last bit of vacant warehouse in our building, we frequently took new and short-term jobs with new and existing customers. Unlike my former employer who boasts a massive finished product warehouse thanks to its facility space and long-term customer relationships, my new employer had no such space and shipped JIT. A pull system instead of a push system. Our customers would tell us what they needed, and we would do what it took to get it to them JIT for them to do their job.

In retrospect, I realize that, just like the product we produced, I was producing our documentation JIT. But I could not continue doing so alone. Since Matt was not assisting with the documentation to the extent I expected, I had to ask for help. I met with Jay, the plant manager, to make my case for an additional writer. He read my mind when he offered up his assistant, Loretta, and agreed to talk to Jerry, the HR manager, to negotiate a temporary consulting position for Hermoso, if he was interested.

MEETING LORETTA

Loretta was a take-charge woman who could make anything happen. She read the QS-9000 manual in less than a week, only asking minimal questions for clarification, and jumped right in assisting with the documentation—from document design to content writing, flowchart design, and inserting images and captions. She was especially adept at determining what the standards were requiring of us. For example, once a form was completed by its user, it became a record per the QS-9000 standards. Given this understanding, Loretta quickly assessed whether we already had records in place to evidence that we were meeting the standards, and if not, she quickly determined the best way to design a form that would provide the data expected of us. It certainly helped that Loretta was a whiz with the computer and the English language.

Her learning curve verified my assumption that it really only takes the patience and wherewithal to read the manual, good thinking and writing skills, and an inquisitive mind to become a QS-9000 specialist. I would joke with my friends, family, and colleagues that I was in demand because I was one of the few who could make it through and understand the tedious manual. Now I was worried that Loretta would realize her external value and leave the company for more lucrative employment opportunities. Perhaps the reams of paper that stood between us and being prepared for the certification audit would prevent her from imagining those possibilities. But I later came to understand, through my subsequent close work with her, that it was her enduring love for Max, the accounting manager who also happened to be her husband, that most likely kept her local. They had met and fallen in love at the company's first factory in Indiana and relocated to Toledo to help start this location together.

HERMOSO'S RESPONSE

With Loretta's reinforcement in place, I took the opportunity to return to the first set of work instructions I drafted from the engineering specification-based SOPs, Hermoso's (and other colleagues') tutorials, and the manufacturing manager's input. Due to the intense information overload I experienced since starting the job, I struggled to remember some of the fine details of the work instructions as I attempted to revise them. Because details are highly important to work instructions—given the safety considerations, customer expectations, and usability—I headed for the floor to confer with Hermoso. I also wanted to gauge his interest in consulting.

I walk out the door to the factory and see Hermoso waiting for me at our usual meeting place, the picnic table in the lunch area. I spread the documentation across the table. We first review the level 1 policy and the level 2 procedure I codrafted with Garrett, the manufacturing manager, line by line, and then we discuss its relationship to the next levels of documentation, the related work instructions and the forms. "The procedure is like the theory that informs the practice, or our instructions, and then the forms are like the output, or evidence, of our daily practices."

Without looking up, Hermoso responds, "I think I will wait to see these other related instructions and forms, then, before I give you feedback on the procedure."

I shuffle the documents around, select two sets of work instructions, place them on top of the stack, fan them out, and start explaining, "SOPs are now called work instructions, and here are two of them that you helped me with in my first couple of weeks. I've finally gotten them in a somewhat coherent format, and I am hoping that you can help me test them out for both comprehension and accuracy."

"Wow, these look different from those SOPs we got hanging out there by our work stations," Hermoso exclaims. "A whole new look. And the text is bigger. That will help us read it better. And the format. I'm not sure I like that. Does it have to be that way?" He points to the tabbing and spacing features that resemble outlines, unlike the SOPs they had been using.

"Not necessarily. Nothing has been approved yet. That's why I am meeting with you. And now is a good time to think about that, too. The Big Three want us to standardize our documentation so there is a certain cohesiveness across our quality manual, procedures, work instructions, and forms. So your feedback early on in this process is so helpful."

At that moment, Hermoso notices both our names as cowriters in a banner across the top of the work instructions. He points to one of them and his face lights up. "Look there. I am a published author."

I take this opportunity to apologize for Matt's appropriation of Hermoso's personalized quality policy without acknowledging his name on the card. "No worries, Angela. I got my personal acknowledgment in the company meeting. That was better than my name on a card that some have already lost, and it will eventually get all bent up in someone's locker, back pocket, or somewhere else anyway. But everyone will remember that day. Or at least I will, and that's all that matters."

"I will remember it, too, Hermoso."

I transition into asking him about collaborating further on QS-9000 documentation, with the possibility of "work release" from the floor to work with me in the front office. He agrees, for which I am thankful, but he wants to strike a daily balance between his window assembly work and his documentation assembly work.

Hermoso and I review copies of the first set of instructions simultaneously. As I read, I circle typos and note my concerns with specific content and formatting in the margins. Hermoso reads in silence. "Well?" I ask, when it appears he has finished.

"I think this is a good start, but I have one question before I give you my opinion: Who are these instructions for? If they are replacing the SOPs, then they should be written for the employees hired to do that job. But if they are written for the Big Three, then that's a whole different story."

I'm taken aback by this response. Why had I not thought about that conflict between these disparate audiences and information needs before? I had successfully guided another company through certification prior to this question. A much larger, more powerful corporation. A corporation without an infrastructure for employee feedback. But it was also a company that was slow to change.

I attempt to answer by thinking it through with him. "The work instructions are technically written for the operator, as the QS-9000 standards just require that we have work instructions for jobs that affect our quality system. But then, again, the Big Three customers specify how they want their jobs done based on their engineer drawings and process descriptions, and the QS-9000 standards have requirements for meeting customer-specific requirements, so I suppose the Big Three still have primary agency over how we write the SOPs. However, perhaps we have primary agency over how we translate those SOPs into work instructions for operator audiences ... and thus your feedback is helpful to making those work instructions better suited to operator audiences."

A NEW APPROACH

This conversation led to several more that subsequently resulted in the development of tactics for making QS-9000 standards work for our company, which meant making our documentation both usable and useful for employees in-house while also meeting the expectations of the Big Three and the independent, external auditor. Moreover, these conversations led me to a deeper personal and organizational understanding of the cultural diversity of our workforce, which ultimately changed the ways in which I worked toward documenting our quality system from then on. The machine operators deserved to have work instructions that represented their diverse ways of making sense of SOPs. Because the operators were the primary audience for these documents, they should be usable and useful for the operators.

One way the documentation could be made more usable and useful for them was to better represent the linguistic and literacy diversity among the operators on the shop floor. Some of the operators spoke, wrote, and read only in English; some spoke, wrote, and read only in Spanish; some only spoke English or Spanish; and for others, there was a good deal of crossover and variance in bilingualism (and even trilingualism, with some of them literate in Nahuatl). Given this new understanding, Hermoso asked a few other employees to collaborate with us on a documentation team that worked toward composing the manufacturing-related work instructions. This team was comprised of Hermoso and a few other native Spanish language speakers and machine operators and a few other native English speakers (including me, the manufacturing manager, and a videographer, who happened to be the product engineer's husband). Accounting for the linguistic and literacy demographics of our workforce, the team developed a unique multigenre and multimediated documentation system that supported and sustained multilingual written and videotaped instructions in both Spanish and English.

This decision resulted in four versions of each work instruction: one print work instruction in English, one print work instruction in Spanish, one video work instruction with English language tutorial (oral, visual, and textual), and one video work instruction with Spanish language tutorial (oral, visual, and textual). Each person on the team contributed to the brainstorming, production, revision, editing, and testing of the instructions, but we each put in more time toward the tasks that spoke to our individual areas of expertise. For example, Hermoso and another colleague took the lead on translating the approved English versions into Spanish, another floor employee served as the voice-over for the Spanish videos, and I helped with the editing of the video tutorials.

As a team, we first developed the work instructions for the SN-95 Mustang quarter-window center-bracket assembly. We conducted usability research and testing early on in the documentation process, which helped us to think through and change some of our content choices. For instance, it was difficult for the videographer to capture the operator simultaneously working on the machine, maintaining good eye contact, and clearly articulating and projecting the instructions for the job. Thus, it often took several cuts and a good deal of editing just to produce a mediocre quality video. Given the feedback on both the process and product, the team decided to employ voice-over tutorials instead, allowing the videographer to focus on capturing the operators doing their typical good job and then later laying an audio track down on the video. The usability feedback on this latter version was much more favorable, due to the improved video quality and clarity of instructional voice. However, even with this improvement, a user identified English signage in the background of the Spanish video during testing, so we later reshot the video for linguistic consistency. We shot the video footage several times and completed numerous rounds of prerelease revisions, but eventually, we completed our first four approved-for-release prototypes, and they were released for circulation. Ironically, despite the numerous changes, these versions were dubbed revision level zero (rev. 0). Each change after this release would require another round of revision, user testing, and approval—and a subsequent release at the next revision level up.

After successfully composing, revising, editing, and testing the first set of SN-95 Mustang quarter-window center-bracket assembly work instructions, our team—in conjunction with human resources—launched the first set of training sessions for both the first- and second-shift SN-95 machine operators. Employees self-selected the media and language version for each training session. The video training in both languages was chosen over the print instructions, which validated our assumption that operators, with their literacy backgrounds and work routine, would have different training preferences than, say, office workers or managers, who spend their days handling written documents. Our multigenre and multimediated documentation system paid off. Per our own newly approved multilingual procedure for assessing training, employees evaluated each training session—either orally (with a translator documenting the feedback) or in writing—and signed their training records.

Once the conference room cleared out following the last session of the day, the team gathered to process our observations of and feedback on the training. The responses to the video training in both languages were overwhelmingly positive from all audiences. From this user feedback, we opted to forgo the print instructions and only implement the video training for these specific work instructions in conjunction with sample assembly prototypes at the work stations (although this decision did not hold consistent for all subsequent assembly instructions). It was then that I realized the documentation system I had once envisioned was no longer just a set of static documents. Instead, it had evolved into a coherent but dynamic and diverse articulation of workplace practices and values.

At the end of a long day of back-to-back consultation, usability testing, training, and assessment meetings, Loretta and I are exhausted and anxious to go home. These weekly marathon sessions are putting our endurance to the test since the first round of usability testing a couple months prior. But with several months remaining before our first external "readiness" audit, we know we must complete our to-do list for the day. Loretta finishes her notes, gathers her things, and calls it a day.

Alone in the conference room, I finish up a few minutes later. As I wearily gather some VHS tapes and quality manuals, Hermoso pokes his head in the door. "Knock, knock," he calls out.

I look up, see that it's Hermoso, and smile. "Hey, Hermoso. What are you doing back here?"

He enters the room carrying two baggies, one in each hand. Each baggie holds six perfectly shaped tamales lined up, snuggled next to one another, each in their own golden corn husk. He holds them up. "I didn't know if you liked spicy food or not, so I brought you six spicy and six mild."

"Oh, Hermoso, gracias! These are absolutely beautiful!"

"That's my name, you know?"

"Pardon?"

"Hermoso. It means beautiful in Spanish."

"I am not surprised."

RECOMMENDED READINGS

1. To learn more about ISO standards and their implications for technical communication and technical communicators, refer to the following sources:

- International Organization for Standardization. (2011). *ISO 9000:2005 – Quality management systems.* Available: http://www.iso.org/iso/iso_catalogue/catalogue_tc/catalogue_detail.htm?csnumber=42180, accessed May 8, 2012.
- International Organization for Standardization document, *Selection and use of the ISO 9000 family of standards.* Available: http://www.iso.org/iso/iso_9000_selection_and_use-2009.pdf, accessed October 14, 2012.
- B. Fisher, "Documenting an ISO 9000 quality system," *Technical Communication,* vol. 42, no. 3, pp. 482–491, 1995.
- K. Schuler, "Preparing for ISO 9000 registration: The role of the technical communicator," in *ACM SIGDOC Proceedings of the 13th Annual International Conference on Systems Documentation.* New York: ACM, pp. 148–154, 1996.

2. To learn more about Gloria Anzaldúa's borderland theories and lived experiences of mestiza consciousness, refer to the following sources:

- G. Anzaldúa, *Borderlands/La Frontera: The New Mestiza,* 3rd ed. San Francisco, CA: Aunt Lute Books, 2007.
- A. L. Keating, *The Gloria Anzaldúa Reader.* Durham, NC: Duke University Press Books, 2009.

3. To learn more about indigenous linguistic, literary, rhetorical, and cultural traditions, refer to the following sources:

- A. Treuer, *Everything You Wanted to Know about Indians But Were Afraid to Ask.* Wadena, MN: Borealis Books, 2012.
- D. H. Justice, *Our Fire Survives the Storm: A Cherokee Literary History.* Minneapolis, MN: University of Minnesota Press, 2006.
- D. Treuer, *Rez Life: An Indian's Journey through Reservation Life.* New York: Atlantic Monthly Press, 2012.
- T. King, *The Truth about Stories: A Native Narrative.* Minneapolis, MN: University of Minnesota Press, 2008.
- M. Mithun, *The Languages of Native North America.* Cambridge: Cambridge University Press, 2001.
- J. Camacho, R. Gutiérrez-Bravo, and L. Sánchez, Eds., *Information Structure in Indigenous Languages of the Americas: Syntactic Approaches.* Berlin: Walter de Gruyter GmbH & Co., 2010.

DISCUSSION QUESTIONS

1. Angela's story shifts between the past tense and the historical present tense (the use of the present tense in narrating past events). How did these shifts make you feel? Did you notice any pattern in the shifts? Why do you think the author deliberately used this rhetoric?

2. What does "subject matter expert" mean? What qualities should they have? Which characters in the stories do you think qualify as subject matter experts? Explain your answers.

3. What particular values and perspectives did Hermoso bring to this QS-9000 project? What might these teach technical communicators who work on intercultural, translation, and/or localization projects?

4. What do you think of the bond between Angela and Hermoso? From where do you think this bond came? Do you think technical communicators should try to develop this kind of bond with the people they work with? If so, how might they do that?

5. What seemed to be the organizational structure and power dynamics at this young, small company in the story? What challenges or opportunities did they create for Angela's, Hermoso's, and other team member's work?

6. Gloria Anzaldúa's hybrid identity is informed by her mestiza consciousness, a state of mind that works in direct response to the colonization of land (specifically, the southwest United States) and the people therein. Given this, consider what might Hermoso's and Angela's hybrid identities be informed by and responding to? What advantages and/or challenges might hybrid identities create in communicating across cultures? What does Angela's story say about this? What are your own experiences and/or thoughts?

7. To create documentations that were useful for the plant operators, Hermoso, Angela, and their team prepared multiple versions of the work instruction to represent the linguistic and literacy diversity of the operators. What do you think of the team's decision and work process? Consider the various contextual factors present in the story when answering this question, including business requirements, industry standardization, *mestiza consciousness*, usability studies, workplace ethics, and/or other factors. Are there some factors that were more important than others? If so, why?

CHAPTER 12

MY LIFE AS AN EFL TRAINER/TECHNICAL TRANSLATOR IN SHANGHAI, CHINA

Huiling Ding

Huiling Ding is an assistant professor in professional communication at North Carolina State University. She worked part time as a corporate trainer, technical translator and interpreter, document designer, and EFL instructor while teaching full time at Tongji University, Shanghai, China, before coming to the United States for her graduate degrees. Her articles have appeared in Technical Communication Quarterly, Written Communication, Business Communication Quarterly, Rhetoric Review, *and* English for Specific Purposes. *She has recently finished a book manuscript on the transcultural communication about the global epidemic of SARS.*

CHAPTER SYNOPSIS

Huiling describes her experience working as an English as a Foreign Language (EFL) teacher/trainer and a contract technical translator/interpreter in the 1990s in China. Her story shows that although technical translation was in high demand, it was looked down upon as drudgery work because the work was hard, the process was messy, and collaboration with team members and clients was nonexistent. Technical translators had little power to speak of, sometimes not even able to secure payment for their hard work. As for technical interpreters, they had to face the challenges caused by office politics and complex power dynamics in the business world. Huiling's story presents these realities, but it also shows that technical translators and interpreters, because of their outsider status and unique skill sets, could assume tremendous power. If technical communication is about enabling and facilitating communication (often on professional and technical topics) between different parties, then in the intercultural context, this work must encompass technical translation and interpretation. Huiling's story invites us to compare the work conditions, power and status, and competences of technical communicators in the United States and technical translators/interpreters in China.

"How many students did you have this morning?" Mr. Li, a talkative, tall and thin man in his early fifties, asked me when we talked outside the empty building for our lunch break.

"Three out of forty," I replied. "What about you?"

"Two out of forty. I was surprised that they came before the New Year's Day of the new millennium," Mr. Li chuckled. Mr. Li and I were the only people left on a high school campus in downtown Shanghai teaching part-time adult students. He taught Japanese and I taught business English. Students would self-select into our class, attend several sessions, and then gradually drop out either because of the high learning curve or because of busy schedules that prevented them from catching up with the class. We would consider it a good class if by the end of the class we still had one-third to half of the students with us.

TEACHING ENGLISH AS A FOREIGN LANGUAGE (EFL)

In the late 1990s, fresh out of college, I worked as a full-time instructor of college English at a comprehensive university in China, teaching the two-course sequence of college English. In my first year, I had quickly learned the teaching materials and found myself with a lot of free time on my hands. Always trying to keep myself busy, I started to look for ways to improve myself as an English teacher and ended up spending a few months studying for the certificate of Advanced Cambridge Business English and Advanced Translation Certificate of Shanghai. I soon passed both tests and found myself, again, with too much free time.

EFL training grew into a booming business in China in the 1990s, generating huge revenues for training corporations and serving a large number of students with different expectations and needs. With both certificates in hand, I soon joined the large group of English professors, ranging from new instructors to full professors, who took part-time jobs teaching. I found a part-time teaching job for a brand new company, EFL Institute, which was associated with a top-ranking university. (All company and character names are pseudonyms.) The owners advertised for part-time instructors in a local newspaper to which I subscribed.

I traveled to downtown Shanghai two evenings a week or two days during the weekend to teach different classes, which I found both stimulating and rewarding.

As I studied for and passed more local and international tests related to English proficiency, I was assigned to teach increasingly difficult classes: translation, interpretation, Cambridge business English, TOEFL, and GRE—anything that could attract large groups of students and that I could prove myself capable of teaching. Sometimes I also got to teach mini–training sessions such as medical English because of my background in English for medical purposes.

Song, the founder of EFL Institute, jokingly told me in one of our early conversations, "We don't need much infrastructure. We hire you, rent a classroom, and recruit students. You teach, we pay you from students' tuition, and if you teach well, we pay you bonus at the end of the year."

MY NEW ROLE AS A TRANSLATOR

Well, I did get a small bonus in the first year. But two years later, the company expanded exponentially: It hired one full-time instructor and hundreds of professors from virtually every university in Shanghai who were now teaching for it on a part-time basis. It expanded its teaching base to five different areas of Shanghai to serve local needs, stopped paying bonuses, and most importantly started a new division on translation and interpretation. Mr. Liu, the cofounder of the institute, assumed the position of manager of the translation division. He faxed all major transnational companies in Shanghai an advertisement highlighting the logo of the top-ranking university with which the institute was supposedly affiliated. And soon enough, the division was flooded with requests for translation from English to Chinese of all kinds of technical documents.

If we define technical communicators broadly as anyone who writes and, in cross-cultural settings, speaks technical content for specific audiences, then technical and business translators and interpreters certainly fit into this category. Many English professors and graduate students with backgrounds in technical or business English would be recruited as technical translators because of the limited supply of such talent. A small group of people with bachelor's degrees in science and technology and good language proficiencies also worked as part-time or full-time freelance technical translators in Shanghai.

After the expansion of the EFL Institute, my role expanded, as did those of quite a few of the institute's part-time instructors who had technical backgrounds. Familiar with both the manager and the general editor in the translation division, we would get recruited, often at the last minute, to work on technical and business projects. Although one did not have to take on the projects, I always considered it a good thing to help whenever possible because of the interesting opportunities they could bring and because of my intention of building a long-term relationship with the company. Ms. Wang, the apologetic secretary, would call me and say, "I am sorry to add more to your workload, Huiling, but all our part-time translators have been assigned projects, and I have no one else with your background to go to. Mr. Lu (the general editor) will owe you a big favor." In one of my trips to teach an English class, I would stop by the office to pick up a thick stack of printed materials to translate, work by myself for a few weeks, and then bring another stack of translated materials with a floppy disk back to the office.

Teaching evening and weekend classes in addition to holding a full-time associate professor position, Mr. Li, like me, would be occasionally recruited to translate technical documents for the translation division, and he reluctantly accepted such assignments. During our lunch in a buffet restaurant that was a five-minute walk away from the campus, Mr. Li told me why he preferred teaching to technical or business translation. "I enjoy regular and flexible teaching schedules and teaching loads. I usually teach two classes on Saturdays only, so I can spend time with my son in the evenings and on Sundays." He paused a little and continued. "They recruit students very well for my classes and they offer much better hourly pay. It offers me a very stable, supplementary income. Besides, I enjoy teaching adult students, who are more mature than college students. Who wants to work long hours staring at a computer screen, translating technical documents from English to Chinese, and getting paid only RMB 50 for every 1000 words when the hourly pay for advanced business English is much better (RMB 80 to 100 for adult classes)?"

With all that said, we both reluctantly accepted translation projects with short due dates that millennium week because Ms. Wang's pool of translators were either working on existing projects or planning to take off the week to celebrate the new millennium.

Having several friends working as full-time or part-time translators, I got introduced to and thus worked for several translation companies at different times. I worked mostly on technical manuals (from English to Chinese) and business documents such as bidding policies, investment guidelines, and import and export letters (from Chinese to English). Documents related to business communication seemed to comprise a main component of the technical translation tasks, partly due to China's increasingly export-oriented economy and its heavy reliance on foreign investments.

I felt quite alienated by the typical roles played by translation companies: brokers of translating projects, avid protectors of identities of existing clients and translators, profit-driven corporations, and employers of temporary, freelance, part-time, and geographically scattered translators. Most translators were graduate students, current or retired English teachers, or people either with a bachelor's degree in English or with good language proficiency and technical background who did not want to or could not find a stable job. Translation companies would publish advertisements for part-time translators, receive many résumés, and pick people with different academic backgrounds for their database of potential translators. When they have to work with new translators, they often send translators some sample passages from current or previous documents. Those who do well in the translation will be hired for smaller projects first.

Often hired by multiple translating companies, most translators work as members of virtual teams without knowing or communicating with one another. In fact, many freelance translators work from home, often living in small cities, telecommuting instead of traveling to their jobs headquartered in larger cities, and arranging their work schedules based on projects. With a large pool of part-time, freelance translators and little quality control in place, technical translation in China suffers from isolated and thus fragmented team products, the common, expedient practices of going with available translators without adequate supervision of the outcomes, and thus inconsistent quality. Moreover, because of the intense competition for translation projects, part-time or full-time freelance translators have little negotiating power in the entire process and may have to accept whatever projects that come their way and the low pay attached to them. They work long hours without health insurance or other benefits.

My friend Alice was in her mid-twenties and worked freelance in Shanghai after graduating from college. She complained to me, "My life is a mess. I work for long hours through nights and weekends in my tiny apartment and eat convenient noodles when my boss gives me projects. My back and hands ache from sitting and typing for weeks. And then I hear nothing from my boss and do nothing for a few weeks. In my unemployed days, I have to calculate how much longer I can pay my mortgage before getting another translation project."

Paid on the total word count of the original document, neither the translators nor the translation agencies have much incentive to strive for better collaboration or teamwork: How can a formula focusing only on precise word counts keep track of individual contribution to collaborative work? How can one feel any motivation to work with team members when projects always come with short deadlines, last-minute, notification, and thick documents full of technical jargon?

At the EFL Institute, when big projects come, Ms. Wang, the secretary for the translation division, would break them into several parts, pick from her long list of translators people with appropriate disciplinary backgrounds, and send files to them separately so that the entire team starts and finishes the projects at the same time. She would never say anything about the larger project or other translators working on the same project, perhaps to protect both the identities of her clients and the hard-to-grow and much cherished list of capable technical translators. One of her favorite instruction was, "You take care of your part, and we will take care of the rest of the project." Therefore, most translators work in isolation and never get connected with the actual clients. Depending on the operating styles of the company and procedures of quality control, big companies may hire full-time editors, like Mr. Lu at the EFL Institute, to edit all translated parts submitted by individual translators to ensure the document's consistency and quality. In contrast, small companies would simply trust whatever their translators produce and ask their secretaries, who may have little language proficiency or editing skills, to put all translated sections together as the final product for their clients.

To make things worse, translators may occasionally find themselves working without pay when small translation companies or individual brokers vanish after getting the final products. Because of the continuously increasing number of part-time and full-time translators in China, the market has become increasingly dominated by the buyers, or translation companies. Clients have a big pool of translators that they can employ for different projects and they get to pick who to work for them and when and how to pay the translators. Many translators experience difficulties negotiating payments. The more aggressive ones may insist on the use of formal contracts or advance payments and drive the client away. In contrast, the less aggressive ones may either work in vain for weeks and never get paid from small-business clients or, if they get lucky, produce good projects and establish an ongoing relationship with that client.

In fact, skipping payment to translators is not that uncommon. Rather than working with translation companies, small clients contract their occasional translation projects to individuals recommended by their employees as competent translators. Depending on the quality of such connections, the quality of translation, and the way negotiations go, the translator may or may not get paid for his or her weeks and months of work. Usually what the translator gets is words of promise of pay. Given China's traditional emphasis on face and mutual trust and given freelance translators' keen competition for projects and the subsequent lack of negotiating power, it is rather rare for translators to ask for formal contracts before starting a project.

Alice once got an offer of a big project online from a translation company that she had never worked with before. She told me about her concerns, "I don't want to alienate them by pushing for a contract, because that shows a lack of trust on my side. And you are supposed to trust them so that they will keep coming back to you for future projects. I will lose a lot of money if they say no and turn to another translator."

I nodded sympathetically and asked, "Can't you at least get some form of promise from them, in an email perhaps?"

She paused a little, frowned, and shook her head, "They use QQ (a popular Chinese instant messaging application) to contact me. There are so many of us competing for jobs now! But I have never worked with them and have no idea how they learned about me. What if they get my translation and never get back to me again?"

After some struggle, Alice decided to accept the project without a written contract. She did take care to finish the first half of the translation, sent it back to the company, and asked for advance payment before moving onto the second part. Eventually it worked out well, and the same company started to send her work on a regular basis afterward.

I had a couple of bad experiences working with individual clients whose promise of pay never materialized. One student from my Cambridge business English class asked me to help with a small translation project for the publishing company for which he worked. He promised to pay the standard market rate after delivery of the final product. I was hesitant at first because I had a pretty busy schedule at that time. However, for several weeks after class, he came to talk to me about that project and asked me to do him a favor since he knew the quality of my work. Eventually, I reluctantly accepted the project and spent a few weekends on it. He disappeared from my class after I gave him all the translation! I called him several times, asking for updates about the project. He would always answer apologetically and say, "I apologize for the delay, Ms. Ding, but we are still waiting for our mother company to approve the budget." After a couple of months, I realized that his repeated promises of payment would never be fulfilled and that it would be a complete waste of my time, so I stopped trying to track him down.

Another project came from a senior colleague of mine at the university where I worked, who coerced several new instructors like me into working on the project for him. When we submitted hard copies of the project to him, what he did was literally put them together and pass the compiled document on to the client. He invited us to a dinner using the pay he got for the project, and he praised the quality of our work, saying, "You folks did an excellent job! I submitted them in their original formats."

MILLENNIUM PROJECT

For the last-minute millennium project pushed on me by EFL Institute, I translated an operational manual for medical equipment from English to Chinese. Ms. Wang picked me because I had a degree in English for medical purposes, even though I did not know much about medical equipment. All I got from her was a thick pile of print paper which contained one-fifth of the original document. Four other part-time translators were working on other parts, I was told, along with the pay for the project: RMB 50 per one thousand words. She sent me on my way with no instruction, no terminology guidelines, nothing.

I spent all my spare time in the next few weeks working on the project. Like other people who worked from home, I enjoyed the unique advantages of working flexibly and without interruption. However, I barely left my apartment except to go to the dining hall for meals or to teach. I slept much less because I wanted to finish the project as quickly as possible so that I could return to my routine. I put all my tools together on my desk: sharpened pencils, blank paper, a medical dictionary and a borrowed engineering dictionary, highlighters of different colors, several floppy disks, and a cup of strong jasmine tea to help me stay awake. My desktop was always on with Internet connections so that I could use online dictionaries if necessary. The speakers were already turned on with light music to keep me from getting bored. I prefer translating first on paper to keep the momentum going and then converting the messy, hand-written product into a clean Word document. It gave me a sense of accomplishment when I typed nonstop and occasionally caught typos in the handwritten translation.

I first highlighted all new words and consulted a medical dictionary and an engineering dictionary for possible meanings. Despite my medical background, the jargon of electronics still posed a high learning curve. Compiling a vocabulary of all the new words together with their Chinese translation was the next task I tackled. Then the translation started. I tried to break down some complicated long sentences into smaller, dynamic ones. Passive voice was converted into active voice, and I consulted friends in electronic engineering when I could not make sense of some passages after putting all the individual meanings of words in those sentences together. When typing my translation as a Word document, I tried to maintain the look of the original document design as much as possible. It was challenging but manageable. The decision to compile a list of new vocabulary helped to streamline the translation and ensured that I used technical terms consistently throughout the document.

However, on several occasions I sat for hours staring at passages I could not make sense of even with the help of my friends. I really wished that I had been given the opportunity to work a little more closely with my unknown and geographically distributed team members, to talk briefly with the actual client who had produced the manual, and to communicate better with both team members and the client to ensure a much smoother translation process and thus a much better product.

I was not the only one who wanted this. When I gave Mr. Lu both the translation and the index of terminology, he seemed quite surprised at my creation and use of additional and "irrelevant" (his original word) materials. "You are the first one to create this list of terms. Why?" He raised his eyebrows after locating the second thin pile of documents with the terminology, "You know we will not count this in your final pay, right?"

I nodded, "I do, but this helped me as a translator, and perhaps you will find it useful as well." He looked suspicious. But he obviously agreed with me after reading all five sections translated by our unconnected team members.

When I dropped by his office after teaching a class on the same floor, he was still proofreading and editing our team product, this time with my compilation of terms spread out all over his desk. "I am so glad that we asked you to work on this," he smiled when I walked in. "Your translation required no editing, and now I am using your index to guide my editing of terms for the other four parts."

"You may as well pay me for creating that list," I said jokingly. "I hope you don't mind my use of short sentences and active voice."

"Too bad I started with your part," Mr. Lu said. "Now that I read this part," he showed me the pages with red marks all over, "I find the long and complex sentences read more like English than Chinese."

I was very familiar with those Chinese complex sentences, which were translated almost literally from their English counterparts. Such clumsy, foreign-looking sentences were all over recently translated literary works.

The kind of quality control Mr. Lu did was not practiced by many of the smaller translation companies for which my friends and I occasionally worked. Driven by budget constraints, instead of hiring professional editors to ensure the correctness and accuracy of the translation, they would hire a secretary to compile different parts of the project together and check only for misspellings and formatting issues. The end product would look exactly like many of the newly translated novels: You would quickly find that they were ghost written, or perhaps I should say, ghost translated, because of the dramatically different styles of individual chapters. Because of this, I had a lot of respect for Mr. Lu for his work, and I volunteered. "I can help to break down those long sentences for you."

OPPORTUNITY FOR INTERPRETATION

Working for the translation division did bring occasional pleasant surprises. Several times I was assigned face-to-face interpretation projects for clients in biomedicine-related fields, which, despite intensive and long working hours, offered much better hourly pay and opportunities to travel. Working on this new millennium project was a stroke of good luck. Just as I was talking with Mr. Lu about those long, clumsy sentences, Ms. Wang interrupted our conversation. "We have no time for editing now. You have to get ready for your interpretation project for a joint venture tomorrow." She passed me a folder with fliers, business cards, and a map. "That's all we know about the project. Good luck!"

Wow, I would work with real clients face to face this time! All I could tell from the package was that the spellings of the two names listed as managers looked like those commonly used in Taiwan or Hong Kong. A little worried about the lack of any information about the actual project, I was relieved at the thought of working directly with people instead of working in isolation on group projects without any knowledge of the team or the client. I did worry a little both about the fact that I, an outsider without any power, would be dropped into an important internal business meeting and about the political tensions such a parachuted outsider might create. Fortunately, many of the students taking my business English and translation classes had full-time jobs in multinational companies, and several of them recommended me as a good EFL trainer to their human resource (HR) managers when training projects came up in their companies. As a result, I had worked as a corporate trainer for top managers in transnational pharmaceutical and electronic companies and gained some experience in the business world.

In all those training projects, I taught business English to senior managers in their offices in the expanding Pudong district across the Huangpu River from the old Shanghai city. I often spent quite a few minutes in the waiting room after an hour-long, corporate-sponsored taxi ride before my students could meet with me for classes. It was often while waiting that I worked carefully with my trainees' grudging administrative assistants, who complained to me about their bosses' exploitation of their work. Suspicious HR managers often viewed my students as competitors, people who boasted national top sales records and were often recently recruited to lead the sales department. HR managers would warn me of my students' strange personality or aggressive working style in the few meetings we had about payment and training plans. Then, during training sessions, my students would complain about their incompetent secretaries and greedy HR managers who knew nothing about sales but still wanted to decide for them whom to hire for their sales teams. What they complained about the most was the glass ceiling created by mother companies in Europe or North America and the lack of understanding or support from foreign CEOs whom they had difficulties communicating with from time to time. Occasionally I was introduced to these foreign CEOs, who could not understand why their top sales managers had requested to work on their English proficiency instead of further improving their sales records.

I once witnessed an office conflict when a student's secretary, Miss Zhu, offended everyone in the company, was in a big fight with the receptionist, and was fired two months after I started my job there. Miss Zhu talked down about her boss often, and once she said to me in the waiting room, "He knows nothing about business contracts, and yet he told me to revise my draft five times today. He simply has no idea what he wants."

I asked, "Really? You worked very hard today!"

She continued, "What makes things worse is that the stupid receptionist does not know how to use the photocopier, and I spent a whole hour fixing it after she tried to copy a document for me." She paused, came nearer, and asked in a loud whisper, "Do you know that she only has a high school diploma and is now studying in an evening college for an associate degree? What a pretty yet useless decorative vase for the company."

In a week, I learned from my student that Miss Zhu was fired after starting a nasty fight with the receptionist. "She is mean spirited, and people thought I had taken her with me when I accepted the offer here. She just happened to leave the same company around the same time I left, and it just makes me angry to think about the assumed connections between her and me," my student concluded after spending five minutes to summarize the entire event for me. Obviously, without Miss Zhu working outside when we had classes, my student felt much more comfortable gossiping about office politics.

I was in a strange power dynamic as a trainer: I was recruited and well respected by HR managers and my students. I worked with them for extended periods of time (usually four to five months), and my students and I negotiated about their learning needs as well as teaching materials and approaches fairly regularly. As an outsider to their companies, I did not participate directly in their business or personal conflicts. But I often offered free taxi rides across the tunnel of Huangpu River to the secretaries and receptionists, young ladies about the same age as I was, and occasionally to HR managers and my students. Then I would ask the driver to drop them off near People's Square so that they could take the subway or another taxi back home. On these cab rides and in those waiting rooms, I became a nonthreatening listener to many of them. I learned to listen, to empathize with whomever I talked with, and to hold my tongue when they complained about one another. Working with all these people with different agendas forced me to navigate through tricky workplace situations and office politics from time to time, which helped greatly whenever I walked into such situations again.

The map Ms. Wang gave me was useless. I called the contact person for the job and was told to take a cab the next morning. I arrived at the joint venture, a food factory, after a 90-minute taxi ride. The receptionist introduced me to Ms. Yang, the HR manager, an energetic woman in her early forties with a Taiwan accent. She was very excited to see me and told me that Mr. Brown, a senior manager from their partner in the United States, would come for a quality control tour and would talk about the financial situation of and future plans for the company. "It is a very important day for the company," she stressed repeatedly, looking rather solemn in her black suit.

After our meeting, she brought me to a conference room with a projector. I sat in the second row, right behind her. "All the head workers in the plant will be here as well," Ms. Yang said. I looked around. Behind a few rows of vacant seats, a couple of dozen workers were talking quietly with a Shanghai accent in the last few rows. I wondered who would sit in those empty rows.

No one did. When a young Chinese man walked in with a middle-aged western-looking man—whom I later knew as Mr. Brown—everyone stopped speaking. Ms. Yang introduced me as the interpreter for the event and the young man as Mr. Zhao, the production manager of the company. Nodding to me slightly, Mr. Zhao sat down in the first row and quickly turned his attention to Mr. Brown, the speaker, without looking at the rest of the audience. Ms. Yang stood up to introduce Mr. Brown in Chinese. I sensed some tension in the air but had no idea from where it came.

It was a strange process. After Ms. Yang's introduction, Mr. Zhao offered an English introduction to welcome Mr. Brown, facing the speaker and standing with his back to his colleagues. Surprised by his apparent lack of interest in his colleagues, I interpreted the entire passage right after the second he stopped for a breath. He seemed a little surprised that I actually did so.

"Why would you spend so much money hiring an interpreter when you have someone in house who is obviously competent enough to do the interpretation?" I wondered.

Then the presentation started. Using several slides, Mr. Brown introduced the financial performance of the mother company and the recently merged plant, then moved on to plans of further development and expansion of the plant. Mr. Zhao talked with him from time to time, ignoring the fact that no one else in the room but Mr. Brown and me could understand what he said. Always focusing his eyes on Mr. Brown, he paid no attention to either Ms. Yang, his partner from Taiwan, or the workers from Shanghai in the room.

"So that may be why he behaved that way. He doesn't even see the need to interpret anything for people in this room," I concluded soon.

A little irritated by Mr. Zhao's arrogance, I tried to intervene early and grasped the pauses Mr. Brown had between slides to interpret the main messages. The strategy seemed to work well against Mr. Zhao's obvious disregard for the large audience in the room. The workers sitting in the back of the room were silent most of the time. But they started to whisper a little among themselves when I interpreted for them the fairly good financial performance of the plant that year and the foreign partner's decision to add more production lines. Ms. Yang looked happy too. When Mr. Brown ended his presentation and asked the room for questions, I emphasized in my interpretation that *everyone* could raise questions. The Q&A session was quite lively with Ms. Yang and a couple of lead workers asking questions and Mr. Brown offering more details about the investment plan and quality control issues.

After the presentation, I went with Mr. Brown and Mr. Zhao for a quality control tour of the production lines in the factory. Two engineers accompanied us once we entered the factory, and Mr. Zhao let me take charge of the translation task in the tour. He listened attentively to what the engineers said about the facilities and joined the conversation from time to time to offer more technical details. Other than Mr. Brown, I was the only one speaking English throughout the process.

Usually the interpretation project ended when I walked out of the client's office. The client will pay the translation division, and I will get my share in a few weeks. However, this project was an anomaly. Ms. Yang called me the next week and invited me to meet with her at a local tea house. She thanked me again for taking charge of the interpretation process and for informing both her and the workers about the presentation. She told me that Mr. Zhao and she were the only two managers sent from their Taiwan headquarters. The Taiwan headquarters had acquired the current plant from a local food factory one-and-a-half years ago and had kept most of the senior workers. Then she and Mr. Zhao got transferred to Shanghai half a year after the merger to take charge of human resources and manufacturing. Mr. Zhao had an undergraduate degree in manufacturing, and she had an associate degree in business management. Ms. Yang had moved to Shanghai a few years ago, leaving all her children in Taiwan. Both she and Mr. Zhao were expected to work in Shanghai for a few years before they could be transferred back to Taiwan, and she missed her family dearly.

"We disagree on numerous things," she spoke quite candidly. "He always thinks that he is the one in charge because he is more proficient in English, has a technical background, and can communicate better with our partners. And he always dominates those presentations," she complained. "He never shares with us what the foreign managers say. I insisted that we hire a professional interpreter this time, and I am so glad that you came."

"So it was about internal politics," I said to myself. Taiwanese companies are often family owned and they have a reputation of underpaying and overworking their employees and offering little mechanism for workers to participate in the management of the companies. The division between Taiwanese managers and mainland workers transferred from an acquired local factory was quite obvious even to outsiders like me, and the subsequent rigid hierarchy between parachuted managers and local workers created a stifling working environment. The personal conflicts between the only two Taiwanese managers certainly made things worse.

I still remember the shock I experienced when I realized how internal politics can not only interfere with interpretation arrangements but also be influenced by temporarily parachuted interpreters. I was invited as an external force to rescue the one-way flow of information from the foreign partner to Mr. Zhao only, and my decisions certainly gave both Ms. Yang and workers the much needed access to the current financial status of and future plans for their plant.

That was not the closure of the project. One year later, Mr. Zhao took the last TOEFL writing class I taught before I left for graduate study in the United States. He looked like a completely different person: He was polite and friendly to his classmates, and he actively participated in our class discussions. Neither of us mentioned the interpretation project during our four-month class, but he brought me a traditional Chinese tea pot and cups for "doing a wonderful job" for his company. He told me that he hoped to pursue his MBA degree in the United States because he saw "no opportunity for further career development" in the company. "What a small world," I thought. "Perhaps we will meet again in future."

AFTERWORD

I maintain my contact with the virtual community of technical translators by visiting the largest national discussion forum for translators in China (http://bbs.transla tors.com.cn), which had 56,200 registered users in 2011. Many participants describe themselves as isolated, telecommuting, freelance or part-time translators with technical or business backgrounds and good English proficiency, and they work as independent entrepreneurs for clients who supply them with projects. A few participants who encountered nonpaying companies started to create a blacklist of companies to avoid for the reference of other technical translators, often providing instant messaging conversation records as the evidence. Currently, the blacklist has a total of 90 companies from all over China. As a grassroots effort, the growing blacklist exerts pressure on translation companies, small companies, and individuals offering translation projects.

One common thing among the forum participants is their active effort in continuous education, particularly in the self-teaching of disciplinary knowledge and in seeking certification in various areas related to translation and interpretation. Several senior participants discuss their experience of spending several years training themselves as experts in specific areas and then finding a niche translation market for their combined expertise in both technical disciplines and disciplinary English. Forum participants in pursuit of advanced certificates exchange information about training courses, study materials, study tactics, and test tips to support one another in their time-consuming yet rewarding endeavors. I wish I could have had access to such a supportive community back in the days instead of working all by myself. Collaboration and crowdsourcing can really change people's lives, particularly the lives of freelance entrepreneurs such as technical translators.

RECOMMENDED READINGS

1. To learn more about the processes and strategies of technical translation, refer to the following sources:

   - U. Ozolins, "Back translation as a means of giving translators a voice," *Translation & Interpreting,* vol. 1, no. 2, pp. 1–13, 2009.
   - D. Walmer, "One company's efforts to improve translation and localization," *Technical Communication,* vol. 46, no.2, pp. 230–237, 1999.
   - J. Fisher, "Improving the usability of online information when translated from English to Chinese," *IEEE Transactions on Professional Communication,* vol. 39, no. 3, pp. 122–128, 1996.
   - K. St.Amant, "Expanding translation use to improve the quality of technical communication," *IEEE Transactions on Professional Communication,* vol. 43, no. 3, pp. 323–326, 2000.
   - M. Cronin, *Translation and Globalization.* New York and London: Routledge, 2003.
   - T. Weiss, "Translation in a borderless world," *Technical Communication Quarterly,* vol. 4, no. 4, pp. 407–425, 1995.

2. Technical translation has experienced important changes as the field of technical communication became increasingly globalized. Consult the following sources to learn more about these changes:

   - M. I. Hallman, "Differentiating technical translation from technical writing," *Technical Communication,* vol. 37, no. 3, pp. 244–247, 1990.
   - C. Séguinot, "Technical writing and translation: Changing with the times," *Journal of Technical Writing and Communication,* vol. 24, no. 3, pp. 285–292, 1994.
   - M. Gnecchi, B. Maylath, B. Mousten, F. Scarpa, and S. Vandepitte, "Field convergence between technical writers and technical translators: Consequences for training institutions," *IEEE Transactions on Professional Communication,* vol. 54, no. 2, pp. 168–184, 2011.

3. For information on how to prepare technical communication students to work with translators or on translation projects, consult the following sources:

- B. Maylath, "Writing globally: Teaching the technical writing student to prepare documents for translation," *Journal of Business and Technical Communication,* vol. 11, no. 3, pp. 339–352, 1997.
- M. Flammia, "Preparing technical communication students to play a role on the translation team," *IEEE Transactions on Professional Communication,* vol. 48, no. 4, pp. 401–412, 2005.
- B. Maylath and E. Thrush, "*Café, thé ou lait?* Training technical communicators to manage translation and localization," in *Managing Global Communication in Science and Technology,* P. Hager and H. J. Scheiber, Eds. New York: Wiley, 2000, pp. 233–254.

DISCUSSION QUESTIONS

1. Compare the work conditions, power and status, and competencies of technical translators/interpreters in China as they are portrayed in Huiling's story with the situation of technical communicators in the United States. Do you see differences? Similarities?

2. Use a currency conversion site to find out the U.S. dollar equivalence of translating technical documents from English to Chinese (RMB 50 for every 1000 words) and teaching advanced business English (RMB 80 to 100 per hour). Remember, however, that these figures reflect China in the late 1990s, so to truly understand the monetary value of these works, one must consider changes in cost of living and currency rates.

3. Collaboration and teamwork are highly valued in technical communication, but working teams often face issues such as nonresponding or free-riding teammates and the lack of motivation and rewarding mechanisms. How do you deal with teamwork problems? What strategies can be employed to best ensure accountability of all team members? What would you do if your team members from different cultures do not share the same value of accountability or simply dislike team work and prefer to work by themselves?

4. Many intercultural technical communicators work as independent entrepreneurs who take charge of their own businesses and expand their clients. What challenges do you think they will encounter in such processes and how might they cope with such challenges?

5. Although Huiling's story told us a great deal about the challenges and difficulties of a contract translator in China, she also suggested that the services offered by translation companies might not have been high quality. What problems in translation services did she mention? Why might such practices be problematic? Discuss possible causes and solutions for these problems.

6. In this story, as in other stories in this collection, we mostly refrained from using "America" to refer to things, people, and so on from the United States and instead used the term "the United States." This is despite the fact that some people, whether residents of the United States or not, actually use "America"

to refer to the United States. Research the linguistic and cultural reasons and significance behind either choice.

7. Do an online search for "technical translation services" and carefully read at least five of the hits you get. How do their services compare with the services offered by the organizations with which Huiling contracted? How do they compare with the services Huiling offered, or tried to offer, to her clients?

8. How did Huiling deal with issues of power? Why was it important in translation work? What did power have to do with linguistic decisions she had to make as a translator? And what did it have to do with her work conditions and relationships?

INDEX

Books in the IEEE PCS
PROFESSIONAL ENGINEERING COMMUNICATION SERIES

Sponsored by IEEE Professional Communication Society

Series Editor: Traci Nathans-Kelly

This series from IEEE's Professional Communication Society addresses professional communication elements, techniques, concerns, and issues. Created for engineers, technicians, academic administration/faculty, students, and technical communicators in related industries, this series meets a need for a targeted set of materials that focus on very real, daily, on-site communication needs. Using examples and expertise gleaned from engineers and their colleagues, this series aims to produce practical resources for today's professionals and pre-professionals.

Information Overload: An International Challenge for Professional Engineers and Technical Communicators · Judith B. Strother, Jan M. Ulijn, and Zohra Fazal (editors and authors)

Negotiating Cultural Encounters: Narrating Intercultural Engineering and Technical Communication · Han Yu and Gerald Savage (editors and authors)

Forthcoming:

Slide Rules: Design, Build, and Archive Presentations in the Engineering and Technical Fields · Traci Nathans-Kelly and Christine G. Nicometo (authors)

Teaching and Training for Global Engineering: Perspectives on Culture and Communication Practices · Kirk St. Amant and Madelyn Flammia (editors and authors)

International Virtual Teams: Engineering Successful Global Communication · Pam Estes Brewer (author)

Printed in the United States
By Bookmasters